Transparency Masters to Accompany
SPC Essentials and Productivity Improvement

Also available from ASQC Quality Press

SPC Essentials and Productivity Improvement: A Manufacturing Approach
William A. Levinson and Frank Tumbelty

Statistical Process Control Methods for Long and Short Runs, Second Edition
Gary K. Griffith

Glossary and Tables for Statistical Quality Control, Third Edition
ASQC Statistics Division

SPC Tools for Everyone
John T. Burr

Quality Control for Operators and Foremen
K. S. Krishnamoorthi

Concepts for R&R Studies
Larry B. Barrentine

To request a complimentary catalog of publications, call 800-248-1946.

Transparency Masters to Accompany
SPC Essentials and Productivity Improvement

A Manufacturing Approach

William A. Levinson

Frank Tumbelty

Harris Corporation, Semiconductor Sector

ASQC Quality Press
Milwaukee, Wisconsin

Transparency Masters to Accompany
SPC Essentials and Productivity Improvement: A Manufacturing Approach
William A. Levinson and Frank Tumbelty

© 1997 by Harris Corporation, Semiconductor Sector

10 9 8 7 6 5 4 3 2 1

ISBN 0-87389-373-5

Acquisitions Editor: Roger Holloway
Project Editor: Jeanne W. Bohn

ASQC Mission: To facilitate continuous improvement and increase customer satisfaction by identifying, communicating, and promoting the use of quality principles, concepts, and technologies; and thereby be recognized throughout the world as the leading authority on, and champion for, quality.

Attention: Schools and Corporations
ASQC Quality Press books, audiotapes, videotapes, and software are available at quantity discounts with bulk purchases for business, educational, or instructional use. For information, please contact ASQC Quality Press at 800-248-1946, or write to ASQC Quality Press, P.O. Box 3005, Milwaukee, WI 53201-3005

For a free copy of the ASQC Quality Press Publications Catalog, including ASQC membership information, call 800-248-1946.

Printed in the United States of America

 Printed on acid-free paper

Quality Press
611 East Wisconsin Avenue
Milwaukee, Wisconsin 53202

Contents

Preface

This set of transparency masters accompanies *SPC Essentials and Productivity Improvement: A Manufacturing Approach.* It is for high school or college instructors, or in-house company trainers. The transparency masters are appropriate for use either in the workplace or in an academic classroom for courses in statistical process control (SPC) and quality improvement.

The basic material in the transparency masters, as in the book, use basic math: addition, subtraction, multiplication, and division. In industry, time costs money. As a practical process management technique, SPC does not require lengthy manual calculations. If the math is not simple enough for rapid calculation by hand, a computer handles it.

We must, however, understand what the computer is telling us about the process. Chapter 3 equips readers to understand and interpret the control charts they may see in the factory. Chapter 2 covers nonmathematical techniques for improving quality and productivity and quality management techniques. It answers questions like, "Why should we use the ISO 9000 standards?"

A basic SPC and quality improvement course should use the transparencies found in chapters 1 through 5, but omit the technical appendices for chapters 4 and 5. An advanced course should add the technical appendices and chapter 6. These sections equip readers to program a computer, or tell commercial SPC software what to do. They show how to characterize processes and set up new control charts for them. This material requires technical mathematics at the advanced high school or freshman college level.

The transparency master set includes the statistical tables that appear in the book appendices. These are for the instructor's reference in the classroom, and for showing students how to use the tables. The instructor should make transparencies of the tables for this purpose.

INTRODUCTION

Transparency Masters to Accompany *SPC Essentials and Productivity Improvement*
ASQC Quality Press   ©1997 by Harris Corporation, Semiconductor Sector

# Introduction: Manufacturing

- ➤ Manufacturing workers—the people who add value to a physical product—play vital roles in creating national wealth.

- ➤ Physical wealth comes only from natural resources and manufacturing.

- ➤ National power, whether economic or military, comes from manufacturing capability. This is the ability to add value to physical products.

Transparency Masters to Accompany *SPC Essentials and Productivity Improvement*
ASQC Quality Press   ©1997 by Harris Corporation, Semiconductor Sector

Quality Is a Competitive Weapon

➤ Quality helps us keep customers and stay in business.

   ➤ Satisfied customers may not even bother looking in the competitor's shop.

➤ Quality can help us take the competitor's business.

➤ Quality can compete against a competitor's lower prices.

➤ Quality is vital in customer retention.

   ➤ It costs five to seven times as much to get a new customer as it does to keep an existing one (Struebing 1996).

Source: Struebing, Laura. 1996. Customer loyalty: Playing for keeps. *Quality Progress* (February): 25.

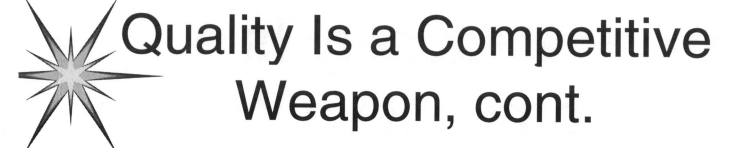

Quality Is a Competitive Weapon, cont.

➤ ## Unhappy customers?

➤ *I command that the owner of the Tula factory Kornila Beloglazov be flogged and banished to hard labor in the monasteries. He, the scoundrel, dared to sell to the Realm's army defective handguns and muskets.*

➤ *Foreman alderman, Frol Fuks should be flogged and banished to Azov, this will teach him otherwise than to put trademarks on faulty muskets.*

—Decree of Tsar Peter I, 11 January, 1723 (Juran 1995, 390–391)

Source: Juran, Joseph. 1995. *A history of managing for quality*. Milwaukee: ASQC Quality Press.

Quality Is a Competitive Weapon, cont.

- ➤ Happy customers provide free advertising.
  - ➤ Unhappy customers bad-mouth our products and services.
- ➤ Impartial quality rating services include magazines such as *Consumer Reports.*
  - ➤ Can advertising overcome a poor rating from *CR*?

Quality Versus Price

➤ Price does not account for all of the customer's costs.

  ➤ Cost of replacing it when it breaks

  ➤ Cost of repairing it when it fails

  ➤ Downtime/unavailability (for example, passenger aircraft, manufacturing equipment)

➤ In manufacturing, poor quality raw materials or parts can make the product more expensive.

What Is Quality?

- ➤ Quality is fitness for use.
    - ➤ The customer defines fitness for use.
    - ➤ A quality product meets or exceeds the customer's requirements.
- ➤ Quality has several aspects.
    - ➤ Customer specifications
    - ➤ Operating costs (for example, fuel economy)
    - ➤ Maintenance costs
    - ➤ Downtime/availability
    - ➤ Safety
    - ➤ Ease of use

Transparency Masters to Accompany *SPC Essentials and Productivity Improvement*
ASQC Quality Press   ©1997 by Harris Corporation, Semiconductor Sector

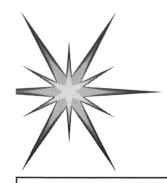

Manufacturing's Role in Quality

- ➤ Manufacturing's primary role is to meet the customer's specifications.
  - ➤ *Statistical process control* (SPC) is a tool for doing this.
  - ➤ Specifications originated because mass production requires interchangeable parts.
    - ➤ Eli Whitney and cotton gin, arms manufacture
- ➤ Good product design will assure other aspects of quality.
  - ➤ This is the product designers' role.

Transparency Masters to Accompany *SPC Essentials and Productivity Improvement*
ASQC Quality Press   ©1997 by Harris Corporation, Semiconductor Sector

Specifications

- ➤ Specification limits are like a football goalpost or a target.

- ➤ Anything inside the limits is good, and anything outside is bad.

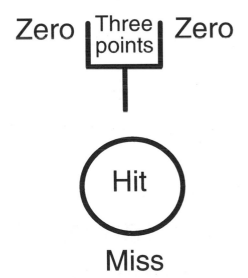

Transparency Masters to Accompany *SPC Essentials and Productivity Improvement*
ASQC Quality Press ©1997 by Harris Corporation, Semiconductor Sector

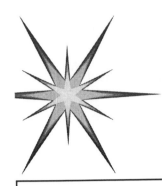

Customers

➤ External customers buy goods or services from our company.

➤ Internal customers are people inside our company to whom we provide a product or service.

➤ All work in process (WIP) passes from an internal supplier to an internal customer.

➤ Meeting the needs of internal customers helps the company satisfy external customers.

Transparency Masters to Accompany *SPC Essentials and Productivity Improvement*
ASQC Quality Press   ©1997 by Harris Corporation, Semiconductor Sector

The Manufacturing Worker's Role in Quality

- ➤ Today's manufacturing worker plays a critical role in assuring and improving quality.

    - ➤ The person who has his or her hands on the materials, equipment, and machine controls learns intimate details about the task.

- ➤ At Harris, self-directed manufacturing teams plan and carry out their own work.

    - ➤ Teams initiate projects to improve productivity and quality.

    - ➤ Engineers, technicians, and others provide technical support.

Transparency Masters to Accompany *SPC Essentials and Productivity Improvement*
ASQC Quality Press   ©1997 by Harris Corporation, Semiconductor Sector

Star Organization for Self-Directed Work Teams

➤ The star organization associates a person with each responsibility.

Production

Safety — **Quality**

Administrative **Team Communications**

Transparency Masters to Accompany *SPC Essentials and Productivity Improvement*
ASQC Quality Press   ©1997 by Harris Corporation, Semiconductor Sector

Customer Contact Teams

➤ Customer contact teams (CCTs) include

  ➤ Primarily manufacturing workers

  ➤ Engineers and managers

➤ They meet with the customer's frontline workers.

➤ The CCT shortens the communication path.

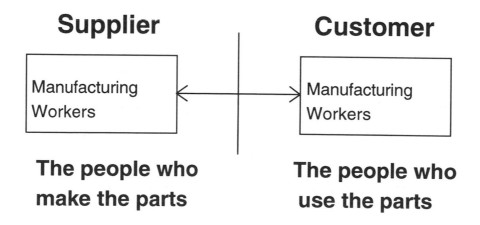

Supplier	Customer
Manufacturing Workers	Manufacturing Workers
The people who make the parts	**The people who use the parts**

Transparency Masters to Accompany *SPC Essentials and Productivity Improvement*
ASQC Quality Press   ©1997 by Harris Corporation, Semiconductor Sector

Advantage of the CCT

➤ This is the traditional customer-supplier communication system.

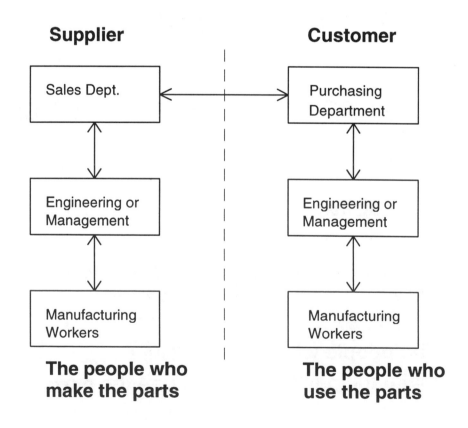

Transparency Masters to Accompany *SPC Essentials and Productivity Improvement*
ASQC Quality Press   ©1997 by Harris Corporation, Semiconductor Sector

Tools for Improving Productivity and Quality

Transparency Masters to Accompany *SPC Essentials and Productivity Improvement*
ASQC Quality Press   ©1997 by Harris Corporation, Semiconductor Sector

Friction

➤ *Friction* refers to seemingly minor annoyances that add up to major problems.

> ➤ We often don't notice them because we can work around them.

> ➤ These minor annoyances and inefficiencies can seriously undermine quality and productivity.

➤ Frontline workers are in the best position to discover and correct sources of friction.

> ➤ The person who does the job eight or more hours a day is in the best position to identify problems.

Transparency Masters to Accompany *SPC Essentials and Productivity Improvement*
ASQC Quality Press   ©1997 by Harris Corporation, Semiconductor Sector

# Origin of Friction

➤ **Then:** "Friction, as we choose to call it, is the force that makes the apparently easy so difficult....Countless minor incidents—the kind you can never really foresee—combine to lower the general level of performance."

—General Carl von Clausewitz, *On War* ([1831] 1976)

➤ **Now:** "The accumulation of little items, each too 'trivial' to trouble the boss with, is a prime cause of miss-the-market delays."

—Tom Peters, *Thriving on Chaos* (1987)

Sources: Clausewitz, Carl von. [1831] 1976. *On war.* Book 1. Translated by M. Howard and P. Paret. Princeton, NJ.; Princeton University Press

Peters, Tom. 1987. *Thriving on chaos.* New York: Harper & Row.

Friction—Examples

➤ "For want of a nail the shoe was lost; for want of a shoe the horse was lost; and for want of a horse the rider was lost." (Benjamin Franklin)

  ➤ The absence or failure of an inexpensive and seemingly unimportant part—a simple nail—has serious consequences.

➤ Fasteners (bolts and capscrews)

  ➤ Fasteners hold airplanes together.

  ➤ Unscrupulous suppliers are selling SAE grade 7 fasteners as grade 8 (the highest grade).

Transparency Masters to Accompany *SPC Essentials and Productivity Improvement*
ASQC Quality Press   ©1997 by Harris Corporation, Semiconductor Sector

Friction—Examples, cont.

- ➤ Gaskets and seals keep corrosive, flammable, and toxic chemicals inside pipes and valves.
    - ➤ Unscrupulous suppliers are putting reputable brand names on counterfeit gaskets and seals.
    - ➤ This is like stamping "Rolex" on cheap watches.
- ➤ The supermarket fails to program a bar code into the price scanner.
    - ➤ The cashier must stop and ask for a price check. This irritates the customers who are waiting in line.

Transparency Masters to Accompany *SPC Essentials and Productivity Improvement*
ASQC Quality Press   ©1997 by Harris Corporation, Semiconductor Sector

Friction—Examples, cont.

- ➤ A bad semiconductor package (transistor cap or stem) jams the packaging machine.

  - ➤ The cap or stem costs about a dime. The machine costs perhaps half a million dollars.

- ➤ Workers are trying to do a job, and they can't find the screwdriver they need.

  - ➤ 5S-CANDO addresses this.

- ➤ A power outage stops the computer and ruins the work.

  - ➤ Uninterruptable power supplies prevent this.

Transparency Masters to Accompany *SPC Essentials and Productivity Improvement*
ASQC Quality Press   ©1997 by Harris Corporation, Semiconductor Sector

Friction: The Hidden Enemy

➤ Friction is insidious because we can often work around it. It becomes a routine part of the job.

   ➤ Fixing breakdowns

   ➤ Reworking parts

   ➤ Sorting (rectifying, detailing) good parts from bad ones

➤ **If it's frustrating, a chronic annoyance, or a chronic inefficiency, it's friction.**

Total Productive Maintenance

➤ *Total productive maintenance* (TPM) is a program for assuring continuity of operations and for reducing defects.

   ➤ Continuity of manufacturing operations means that the equipment is available when we need it.

Life Cycle Cost for Equipment

- ➤ This is the real cost of buying and owning a durable item.
  - ➤ Purchase price
  - ➤ Operating costs (fuel economy, power efficiency)
  - ➤ Repair and maintenance
  - ➤ Insurance costs
  - ➤ Disposal costs
- ➤ The purchase price may be only a small part of the cost of owning the item.
  - ➤ It may be better to pay more up front to buy reliable equipment.

Opportunity Costs

- ➤ This is the cost of missing the opportunity to make money.

- ➤ A breakdown in a passenger airplane costs $2000 to fix. This is a repair cost.

  - ➤ The flight would have made a $5000 profit. This is the opportunity cost.

    - ➤ The accounting system reports the $2000, but not the $5000.

  - ➤ The airplane breakdown will probably inconvenience the passengers, and they may choose a competitor the next time. This is an intangible cost.

Transparency Masters to Accompany *SPC Essentials and Productivity Improvement*
ASQC Quality Press   ©1997 by Harris Corporation, Semiconductor Sector

"Cheap Is Dear"

- ➤ Two machines will do a job. Annualized purchase costs are
  - ➤ Machine A, $20,000
  - ➤ Machine B, $30,000
- ➤ Annual operating costs
  - ➤ Machine A, $7000
  - ➤ Machine B, $4000
- ➤ Machines earn $500/day.
- ➤ Downtime
  - ➤ A, 22 days/year
  - ➤ B, 2 days/year

Transparency Masters to Accompany *SPC Essentials and Productivity Improvement*
ASQC Quality Press   ©1997 by Harris Corporation, Semiconductor Sector

"Cheap Is Dear," cont.

➤ Which machine is better?

	Machine A	Machine B
Purchase cost (annualized)	$20,000	$30,000
Life cycle costs, excluding the purchase cost	$7000	$4000
Opportunity costs	$11,000	$1000
Loss of customer goodwill	22x	2x
Total	$38,000 + 22x	$35,000 + 2x

Transparency Masters to Accompany *SPC Essentials and Productivity Improvement*
ASQC Quality Press   ©1997 by Harris Corporation, Semiconductor Sector

# An Open Door for a Competitor

➤ Suppose that we have an equipment breakdown and cannot meet a delivery schedule.

  ➤ The customer who was relying on us buys Brand X instead.

  ➤ This is the first time the customer has done business with Brand X.

  ➤ Our failure to meet the delivery schedule lets Brand X "get its nose under the tent."

  ➤ Remember that consistent satisfaction may keep the customer from looking in the competitor's shop.

Transparency Masters to Accompany *SPC Essentials and Productivity Improvement*
ASQC Quality Press   ©1997 by Harris Corporation, Semiconductor Sector

Workstation Availability

$$\text{Availability} = \frac{\text{Uptime}}{\text{Uptime} + \text{Downtime}} \times 100\%$$

Uptime	Downtime
• The unit is operating. • The unit is idle, but available for use.	• Under repair • Waiting for parts • Waiting for paperwork

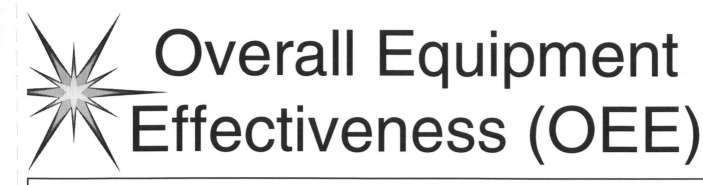

Overall Equipment Effectiveness (OEE)

➤ OEE measures the net effectiveness of a manufacturing tool.

OEE =

Availability ×

Operating Efficiency ×

Rate Efficiency ×

Rate of Quality

OEE, cont.

The availability is the equipment's percentage of uptime.	$$\frac{\text{Uptime} = \text{operating time} + \text{idle time}}{\text{Uptime} + \text{Downtime}} \times 100\%$$
The operating efficiency reflects the idle time portion of the uptime.	$$\frac{\text{Operating time}}{\text{Uptime} = \text{operating time} + \text{idle time}}$$ $$\times 100\%$$
The rate efficiency is the ratio of the unit's actual output to its theoretical output.	$$\frac{\text{Actual output (pieces / time)}}{\text{Theoretical output (pieces / time)}} \times 100\%$$
The rate of quality is the percentage of the output that is good.	$$\frac{\text{Good pieces}}{\text{Actual output} = \text{good and bad pieces}}$$ $$\times 100\%$$

Transparency Masters to Accompany *SPC Essentials and Productivity Improvement*
ASQC Quality Press   ©1997 by Harris Corporation, Semiconductor Sector

OEE, cont.

Simplified expression for OEE

$$OEE = \frac{\text{Operating time}}{\text{Total time}} \times \frac{\text{Good pieces}}{\text{Theoretical output}} \times 100\%$$

Example: Here are data for a machine for a 24-hour day.

Downtime	2 hours
Idle time	3 hours
Theoretical rate	100 pieces/hour
Actual output	1615 pieces
Rework or scrap	97 pieces

Transparency Masters to Accompany *SPC Essentials and Productivity Improvement*
ASQC Quality Press   ©1997 by Harris Corporation, Semiconductor Sector

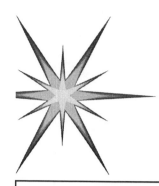

OEE Example, cont.

Availability	$\dfrac{(24-2 \text{ downtime}) \text{ hours}}{24 \text{ hours}}$ $\times 100\%$	91.7%
Operating efficiency	$\dfrac{(22 \text{ uptime} - 3 \text{ idle}) \text{ hours}}{22 \text{ hours}}$	86.4%
Rate efficiency	$\dfrac{1615 \text{ pieces}}{19 \text{ hours} \times \dfrac{100 \text{ pieces}}{\text{hour}}}$	85.0%
Rate of quality	$\dfrac{(1615-97) \text{ good pieces}}{1615 \text{ pieces}}$	94.0%
OEE	$\dfrac{19 \text{ hours}}{24 \text{ hours}} \times \dfrac{1518 \text{ good pieces}}{19 \text{ hours} \times \dfrac{100 \text{ pieces}}{\text{hour}}}$	63.3%

Transparency Masters to Accompany *SPC Essentials and Productivity Improvement*
ASQC Quality Press ©1997 by Harris Corporation, Semiconductor Sector

OEE Example, cont.

- ➤ The machine's effectiveness is 63% of what it would be if it always operated at full speed and made no rework or scrap.

- ➤ Two machines with 95% OEEs can do the work of three machines with 63% OEEs.

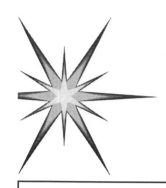

Applicability of OEE

➤ Operating and rate efficiencies are critical only for a constraint or bottleneck process.

  ➤ The constraint or bottleneck is the slowest operation in the process, and it limits the overall productivity.

  ➤ Except at the bottleneck, it is acceptable to have idle time and partial loads.

Preventive Maintenance

➤ Idle time and partial loads are acceptable at nonbottleneck operations.

  ➤ Rework and scrap are never desirable.

  ➤ Equipment breakdowns are never desirable, since it takes time and money to correct them.

➤ Prevention is almost always cheaper than correction.

  ➤ Change the oil in a car.

  ➤ Check belts and hoses in a car.

  ➤ Get a vaccination.

Transparency Masters to Accompany *SPC Essentials and Productivity Improvement*
ASQC Quality Press   ©1997 by Harris Corporation, Semiconductor Sector

Five Major Elements of TPM

1. Improvement activities for making equipment more effective reduce

  ➤ Breakdowns

  ➤ Setups and adjustments

  ➤ Idling and minor stoppages

  ➤ Speed losses

  ➤ Defects, rework, and scrap

  ➤ Startup losses (those due to a setup change or machine adjustment)

Source: Shirose, Kunio. 1992. *TPM for operators*. Cambridge, Mass.: Productivity Press, p. 11.

Five Major Elements of TPM, cont.

2. Autonomous maintenance by manufacturing operators

- Manufacturing operators perform routine maintenance activities such as cleaning, inspection, and lubrication.
- They also learn to recognize and respond to abnormal conditions.

3. Planned maintenance

- These activities are similar to the maintenance schedule for an automobile.

Source: Shirose, Kunio. 1992. *TPM for operators*. Cambridge, Mass.: Productivity Press, p. 11.

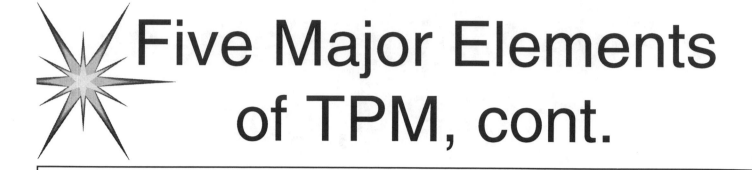

Five Major Elements of TPM, cont.

4. Training to improve operation and maintenance skills

> ➤ Manufacturing operators learn to perform routine maintenance tasks (autonomous maintenance) and identify abnormalities.

> ➤ Maintenance workers and machine attendants learn advanced skills and techniques.

5. Design for maintenance prevention (MP)

> ➤ Design new equipment to reduce its maintenance needs, which makes preventive maintenance and repair easier.

Source: Shirose, Kunio. 1992. *TPM for operators*. Cambridge, Mass.: Productivity Press, p. 11.

TPM and 5S-CANDO

- ➤ 5S-CANDO is a systematic program for cleaning and organizing the workplace.

  - ➤ Systematic cleaning helps prevent breakdowns and defects.

  - ➤ Cleaning equipment provides an opportunity to inspect it.

  - ➤ A clean workplace makes it easy to detect abnormalities.

- ➤ 5S-CANDO is a set of activities for reducing friction and making the workplace safe.

Transparency Masters to Accompany *SPC Essentials and Productivity Improvement*
ASQC Quality Press   ©1997 by Harris Corporation, Semiconductor Sector

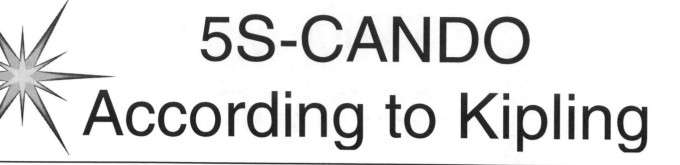

5S-CANDO
According to Kipling

➤ Rudyard Kipling's poem, *The 'Eathen*, says to "Keep your rifle and yourself just so."

   ➤ Cleaning and maintenance of equipment and the workplace is systematic and routine.

   ➤ It shows how to rule the world (or the marketplace).

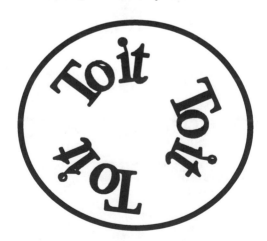

Transparency Masters to Accompany *SPC Essentials and Productivity Improvement*
ASQC Quality Press   ©1997 by Harris Corporation, Semiconductor Sector

# Get a Round to It

➤ The barbarians maintained their equipment "when they got around to it."

- ➤ The English themselves were once the barbarians.

- ➤ Their clothing consisted of woad (blue dye): "one size fits all."

- ➤ "I'll sharpen my spear when I get around to it." (Round to it = A sharpening stone?)

➤ The Romans kept their swords, armor, and themselves "just so."

- ➤ This is why they ruled everything south of Hadrian's Wall.

A Round to It, cont.

➤ 5S-CANDO can help us rule the marketplace.

Japanese 5S		CANDO
Seiri	Clearing up	Clearing up
Seiton	Organizing	Arranging
Seiso	Cleaning	Neatness
Shitsuke	Discipline	Discipline
Seiketsu	Standardization	
		Ongoing Improvement

Transparency Masters to Accompany *SPC Essentials and Productivity Improvement*
ASQC Quality Press   ©1997 by Harris Corporation, Semiconductor Sector

Elements of 5S-CANDO: (1) Clearing Up

➤ Remove nonessential items from the work area.

  ➤ Develop a system to prevent nonessential items from collecting.

  ➤ Put red tags on items that are believed to be nonessential.

  ➤ Put red-tagged items into a holding area. If no one claims them after a given time, discard them.

Source: "Can Do Workshop," Harris Semiconductor, Mountaintop.

Clearing Up, cont.

➤ Organize items by frequency of use.

  ➤ Frequent (daily) use; keep at the workstation.

  ➤ Average (weekly) use; keep near the workstation.

  ➤ Infrequent (monthly) use; remove from the area.

(2) Arrangement

- ➤ "A place for everything, and everything in its place."
  - ➤ Develop an efficient storage system. Get appropriate cabinets, racks, boxes, and so on.
  - ➤ Label tools, equipment, parts, and their storage locations.
  - ➤ Make items easy to remove and return.
    - ➤ Example: Hang wrenches on a wall rack with a place for each wrench.

(3) Neatness

➤ Keep the process equipment, tools, walls, and shop floor clean.

  ➤ Many abnormalities and problems will reveal themselves.

    ➤ Example: A pipe has a slow leak. If the leak is over a clean floor, it will be quickly found.

➤ Remove dirt and debris that can interfere with the equipment or damage the product.

  ➤ In semiconductor manufacturing, even microscopic dust particles can ruin the product.

Transparency Masters to Accompany *SPC Essentials and Productivity Improvement*
ASQC Quality Press   ©1997 by Harris Corporation, Semiconductor Sector

(4) Discipline

- Standardize operations and activities, and follow the standards.
  - Everyone receives appropriate training.
  - Employees help create the rules, and create and modify checklists.
  - Management shows commitment to CANDO.
  - Daily checking and cleaning become routine.

Transparency Masters to Accompany *SPC Essentials and Productivity Improvement*
ASQC Quality Press   ©1997 by Harris Corporation, Semiconductor Sector

(5) Ongoing Improvement

➤ Generate and adopt new and innovative ideas and aggressive goals.

➤ Do not accept abnormalities (friction).

   ➤ Do not let friction become a routine part of the job by working around it.

Synchronous Flow Manufacturing

➤ A factory exists to make money. It does this only by producing and selling finished goods.

  ➤ Operating efficiencies support this goal only if they improve throughput.

  ➤ It is a mistake to produce inventory solely to maintain operating efficiencies.

Transparency Masters to Accompany *SPC Essentials and Productivity Improvement*
ASQC Quality Press   ©1997 by Harris Corporation, Semiconductor Sector

Misuse of OEE

- ➤ A three-step process makes a product. The company uses overall equipment effectiveness (OEE) to measure the work teams' performance.

- ➤ Suppose we run each workstation as fast as possible, to maximize efficiency.

  - ➤ Situation after one hour

Work in process (WIP) waiting for station 2

Station 1 15/hour				Station 2 12/hour		Station 3 20/hour

Misuse of OEE, cont.

➤ After four hours

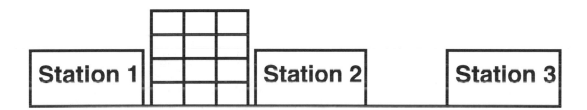

| Station 1 | | Station 2 | Station 3 |

➤ After eight hours (one shift)

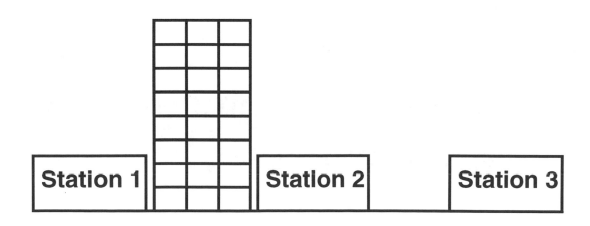

| Station 1 | | Station 2 | Station 3 |

Transparency Masters to Accompany *SPC Essentials and Productivity Improvement*
ASQC Quality Press   ©1997 by Harris Corporation, Semiconductor Sector

Misuse of OEE, cont.

➤ After a year or two

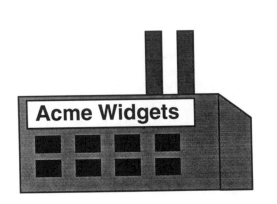

Acme Widgets

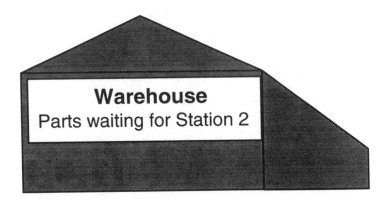

Warehouse
Parts waiting for Station 2

➤ The problem is that station 1 can make 15 pieces an hour, but station 2 can process only 12 pieces an hour.

Is Inventory an Asset?

- ➤ Cost accounting systems treat inventory as a current asset.
  - ➤ It is, however, worth nothing until we sell it. Bills cannot be paid with inventory.
    - ➤ Cash is a liquid asset.
    - ➤ Liquidity ratio acid test = Liquid assets / Current liabilities.
  - ➤ Inventory ties up cash that we could use for other purposes.
    - ➤ Cash flow is often critical in business.

Transparency Masters to Accompany *SPC Essentials and Productivity Improvement*
ASQC Quality Press   ©1997 by Harris Corporation, Semiconductor Sector

The Constraint or Bottleneck

➤ A multistep manufacturing process cannot work faster than its slowest operation or constraint.

> ➤ Trying to make it go faster simply generates piles of inventory in front of the constraint.

➤ *Synchronous Flow Manufacturing* (SFM) is a technique for reducing inventory.

> ➤ SFM ties production starts to the constraint. The constraint sets the pace for the entire process.

Transparency Masters to Accompany *SPC Essentials and Productivity Improvement*
ASQC Quality Press   ©1997 by Harris Corporation, Semiconductor Sector

Synchronous Flow Manufacturing

➤ In the example, station 1 will normally make 12 pieces per hour.

   ➤ This limits its OEE to 80% (12 divided by 15). Don't worry about it.

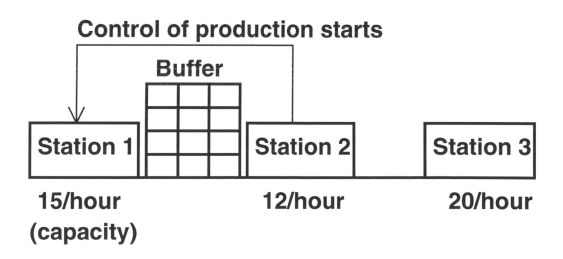

Control of production starts

Buffer

| Station 1 | | Station 2 | Station 3 |

15/hour (capacity)   **12/hour**   **20/hour**

Transparency Masters to Accompany *SPC Essentials and Productivity Improvement*
ASQC Quality Press   ©1997 by Harris Corporation, Semiconductor Sector

SFM, cont.

➤ Time losses at the constraint are irrecoverable.

    ➤ Other operations can make up lost time by working faster.

    ➤ The inventory buffer allows station 2 (the constraint) to keep working even if station 1 breaks down.

Transparency Masters to Accompany *SPC Essentials and Productivity Improvement*
ASQC Quality Press   ©1997 by Harris Corporation, Semiconductor Sector

SFM and OEE

- ➤ We saw that an 80% OEE for station 1 is not only acceptable, but desirable.
  - ➤ If >80%, the inventory pile will keep growing.
- ➤ Station 3's OEE cannot exceed 60%.
  - ➤ 12/hour from station 2 ÷ 20/hour capacity

SFM and OEE, cont.

- ➤ Don't worry about OEE, except at bottlenecks.

    - ➤ This applies to partial loads and idle time.

    - ➤ Downtime due to breakdowns, and inefficiencies due to rework and scrap, are never desirable.

- ➤ Scrap in or after the bottleneck is irreplaceable.

    - ➤ Opportunity cost is the loss of an opportunity to sell the product to a customer. This is a foregone profit not reflected by the accounting system.

Transparency Masters to Accompany *SPC Essentials and Productivity Improvement*
ASQC Quality Press   ©1997 by Harris Corporation, Semiconductor Sector

# SFM and Rework

➤ Rework in the bottleneck is almost as bad as scrap after it.

➤ This also loses the capacity to make a unit.

➤ It can be worse than scrap before the bottleneck!

➤ Scrap that happens before the constraint is replaceable.

SFM, Rework, and Scrap (Overview)

	Rework	Scrap
Before the constraint	Recoverable	Recoverable
At the constraint	Irrecoverable	Irrecoverable
After the constraint	Recoverable	Irrecoverable

Transparency Masters to Accompany *SPC Essentials and Productivity Improvement*
ASQC Quality Press   ©1997 by Harris Corporation, Semiconductor Sector

SFM and Costs: Example

➤ Assume that the process has the following labor and material costs.

➤ Station 2 is the constraint.

Each station's processing cost: $1

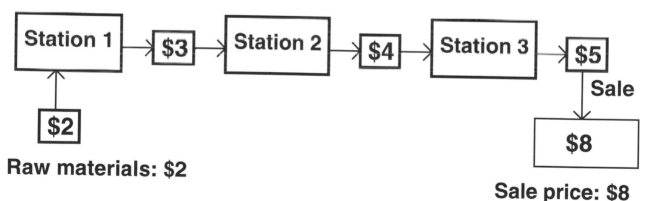

Raw materials: $2

Sale price: $8

$3 Accumulated labor and materials

Transparency Masters to Accompany *SPC Essentials and Productivity Improvement*
ASQC Quality Press   ©1997 by Harris Corporation, Semiconductor Sector

Scrap Loss After Nonconstraint

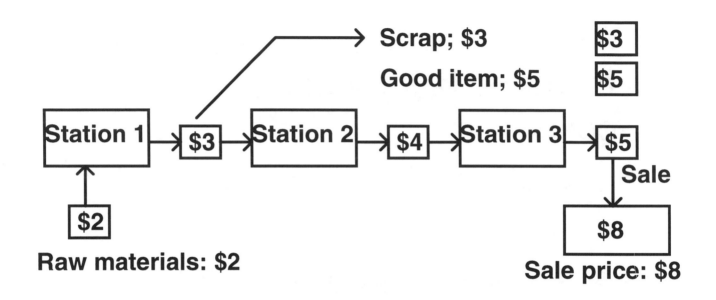

Scrap; $3

Good item; $5

$3

$5

Station 1 — $3 → Station 2 — $4 → Station 3 — $5

$2

Raw materials: $2

Sale

$8

Sale price: $8

➤ Scrap after station 1

   ➤ $3 for scrapped piece

   ➤ $5 to replace it

   ➤ $8 selling price

   ➤ Net gain/loss = $0

Rewrok at Constraint

2 units of capacity ==> 1 rework + 1 piece

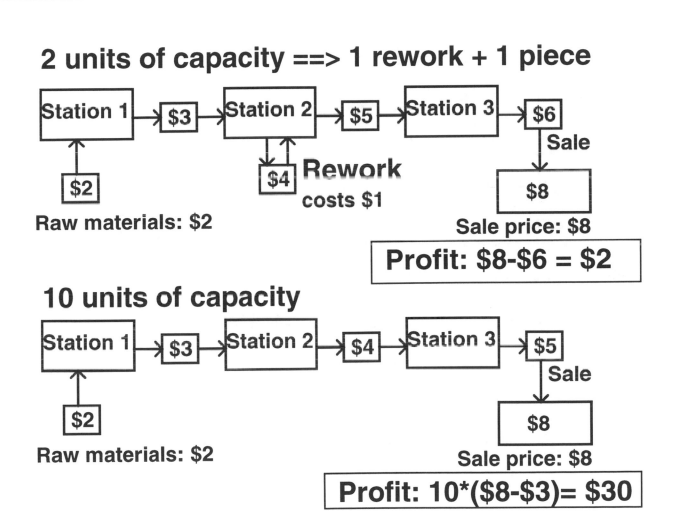

Station 1 → $3 → Station 2 → $5 → Station 3 → $6

$2

Rework costs $1 ($4)

Sale → $8

Raw materials: $2

Sale price: $8

Profit: $8-$6 = $2

10 units of capacity

Station 1 → $3 → Station 2 → $4 → Station 3 → $5

$2

Sale → $8

Raw materials: $2

Sale price: $8

Profit: 10*($8-$3)= $30

➤ In one hour, we make $32.

Transparency Masters to Accompany *SPC Essentials and Productivity Improvement*
ASQC Quality Press   ©1997 by Harris Corporation, Semiconductor Sector

Rework at Constraint, cont.

- ➤ With a rework at the constraint, we made $32 in one hour.
  - ➤ We should have made $36 (12 pieces × $3 profit on each).
  - ➤ The rework cost $1 in processing costs.
  - ➤ $36 – $32 = **$4 loss**, not $1. Why?
    - ➤ $1 to process the rework
    - ➤ $3 for the missed opportunity to make a unit for $5 and sell it for $8

Transparency Masters to Accompany *SPC Essentials and Productivity Improvement*
ASQC Quality Press   ©1997 by Harris Corporation, Semiconductor Sector

SFM, Rework, and Scrap: Conclusion

- ➤ Scrap before the constraint is replaceable.

- ➤ Scrap in or after the constraint is irrecoverable.

- ➤ Rework in the constraint is irrecoverable.

  - ➤ Rework at other operations does not limit the factory's capacity.

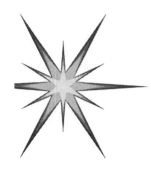

Check Sheets

➤ *Check sheets* or *tally sheets* are simple tools for answering the question, "How often does something happen?"

   ➤ How often does a measurement fall in a certain range?

   ➤ How many pieces are rework or scrap, and why?

Transparency Masters to Accompany *SPC Essentials and Productivity Improvement*
ASQC Quality Press   ©1997 by Harris Corporation, Semiconductor Sector

Check Sheet: Example

➤ Check sheet for semiconductor manufacturing reworks

Rework cause	October 1996				
	6	7	8	9	Total
Bridging metal	ⅢⅠ	Ⅲ	Ⅲ ⅢⅠ	Ⅲ Ⅱ	26
Misalignment	Ⅰ	Ⅲ	Ⅲ	Ⅱ	9
Contamination	Ⅲ	Ⅰ	Ⅰ	Ⅰ	8
Poor contact	Ⅲ Ⅲ Ⅲ	Ⅲ Ⅲ	Ⅲ Ⅲ Ⅱ	Ⅲ Ⅲ Ⅲ	52
Poor development	Ⅱ	Ⅲ	Ⅰ	Ⅰ	7
Total reworks	29	22	25	26	102

Transparency Masters to Accompany *SPC Essentials and Productivity Improvement*
ASQC Quality Press   ©1997 by Harris Corporation, Semiconductor Sector

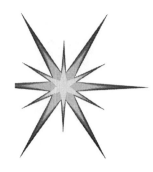

Check Sheet: Example, cont.

- ➤ Poor contact accounts for more than half of the reworks.
  - ➤ A *Pareto chart* shows graphically that this is the worst problem.
  - ➤ A Pareto chart is a special *histogram*. A histogram shows how often something happens.

- ➤ There were five reworks for contamination on October 6.
  - ➤ Is this unusually high?
  - ➤ An *attribute control chart* tells us if defects, rework, or scrap are unusually high.
  - ➤ A *multiple attribute control chart t* tells us if a particular cause is making too many defects or rejections.

Transparency Masters to Accompany *SPC Essentials and Productivity Improvement*
ASQC Quality Press ©1997 by Harris Corporation, Semiconductor Sector

Histograms

➤ Here are the totals from the check sheet example.

➤ Rotate this figure 90° counterclockwise.

Rework cause		Total
Bridging metal	||||||||||||||||||||||||||	26
Misalignment	|||||||||	9
Contamination	||||||||	8
Poor contact	|||||||||||||||||||||||||||||||||||||| |||||||||||||	52
Poor development	||||||||	7

Transparency Masters to Accompany *SPC Essentials and Productivity Improvement*
ASQC Quality Press   ©1997 by Harris Corporation, Semiconductor Sector

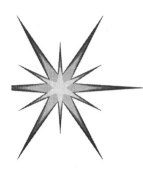

Histograms, cont.

➤ Here is the rework information, in histogram form.

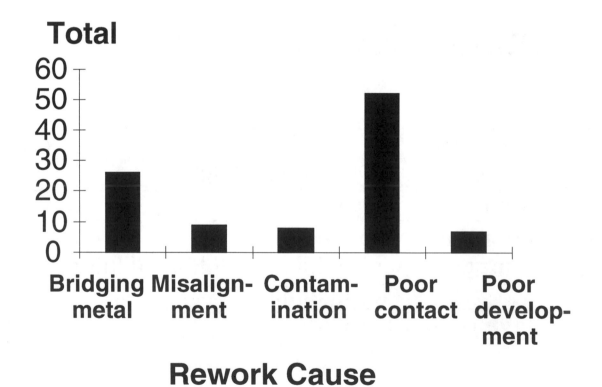

Pareto Chart

➤ The histogram shows how often something happens.

➤ The Pareto chart presents the same information in order of frequency.

  ➤ This helps prioritize efforts to remove problems or improve quality.

Transparency Masters to Accompany *SPC Essentials and Productivity Improvement*
ASQC Quality Press   ©1997 by Harris Corporation, Semiconductor Sector

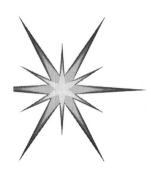

Pareto Chart, cont.

➤ Here is the rework histogram in Pareto chart form.

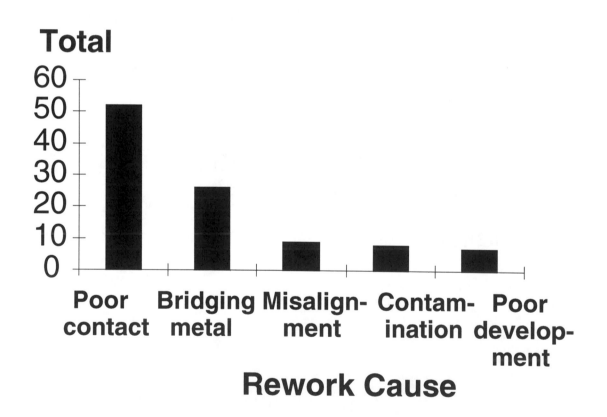

Transparency Masters to Accompany *SPC Essentials and Productivity Improvement*
ASQC Quality Press   ©1997 by Harris Corporation, Semiconductor Sector

Application of the Pareto Chart

- ➤ The Pareto chart is a tool for ranking and sorting alternatives.
  - ➤ It helps prioritize manufacturing problems for attention.
  - ➤ The Pareto Principle says that most of our trouble comes from a few causes.
    - ➤ General rule: 80% of the trouble comes from 20% of the sources.
  - ➤ "Three strikes" laws suggest the Pareto Principle.
    - ➤ Pareto Principle: A few criminals commit most of the crime.
    - ➤ Defects, rework, and scrap are the "criminals" that vandalize our work.

Transparency Masters to Accompany *SPC Essentials and Productivity Improvement*
ASQC Quality Press   ©1997 by Harris Corporation, Semiconductor Sector

# Data Collection for Pareto Charting

1. Use real-time observation.

   ➤ Record events as they occur. Use a check sheet or tally sheet.

2. Use historical data.

   ➤ Caution: Old records have less relevance to the current situation.

3. Perform random sampling.

Transparency Masters to Accompany *SPC Essentials and Productivity Improvement*
ASQC Quality Press   ©1997 by Harris Corporation, Semiconductor Sector

Histograms and Measurements

➤ Histograms can also show the frequency of numerical measurements.

    ➤ These histograms are important in SPC. They show the measurement's distribution.

Measurement Histogram: Example

➤ Frequency of silicon dioxide measurements, in angstroms.

(1) Cells	(2) Count	(3) Total
up to 980	I	1
980 to 985	II	2
985 to 990	IIIIII	6
990 to 995	IIIIIIII	8
995 to 1000	IIIIIIII	8
1000 to 1005	IIIIIIII	8
1005 to 1010	IIIIIIIII	9
1010 to 1015	IIIII	5
1015 or more	III	3

Transparency Masters to Accompany *SPC Essentials and Productivity Improvement*
ASQC Quality Press   ©1997 by Harris Corporation, Semiconductor Sector

Measurement Histogram, cont.

➤ Here is the histogram of the same data.

Frequency

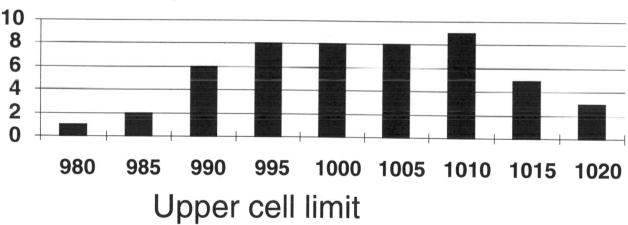

Upper cell limit

Transparency Masters to Accompany *SPC Essentials and Productivity Improvement*
ASQC Quality Press   ©1997 by Harris Corporation, Semiconductor Sector

Measurement Histogram, cont.

➤ Too few or too many cells will yield a histogram that does not show the data distribution well.

  ➤ A good starting rule is to use the square root of the data count.

  ➤ The square root of 50 is about 7. Here is the histogram with 7 cells.

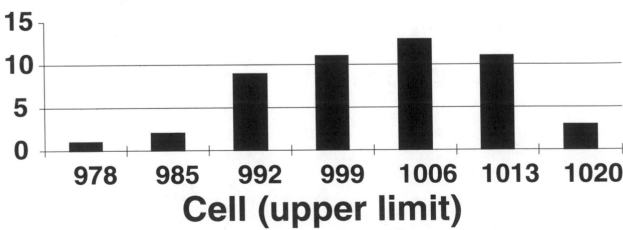

Frequency

Cell (upper limit): 978, 985, 992, 999, 1006, 1013, 1020

Transparency Masters to Accompany *SPC Essentials and Productivity Improvement*
ASQC Quality Press   ©1997 by Harris Corporation, Semiconductor Sector

Histogram: Application

➤ The SPC section will show that data from most manufacturing processes follow a *normal distribution* or *bell curve*.

　➤ The histogram's shape should be similar to a bell.

Transparency Masters to Accompany *SPC Essentials and Productivity Improvement*
ASQC Quality Press　　　　　　©1997 by Harris Corporation, Semiconductor Sector

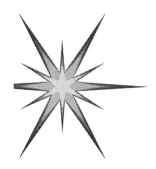

Process Flowcharts

➤ A *flowchart* shows the steps in a manufacturing process. It is useful for

    ➤ Understanding the process

    ➤ Identifying critical operations

Process Flowcharts: Typical Symbols

Operation (product-ion or activity)	◯	Flow of work	⟶
100% inspection, testing, or measure-ment	▢	Inspect-ion or measure-ment, with a control chart	▩

➤ These are only examples. Different factories may use different symbols.

Transparency Masters to Accompany *SPC Essentials and Productivity Improvement*
ASQC Quality Press   ©1997 by Harris Corporation, Semiconductor Sector

Process Flowcharts, cont.

➤ A refinement of the chart can show where materials enter the process.

- ➤ Raw materials or subassemblies can affect product quality or productivity.

- ➤ Showing where they enter the process can help isolate problem sources.

➤ Another refinement shows an operation block for each workstation that can perform the operation.

Transparency Masters to Accompany *SPC Essentials and Productivity Improvement*
ASQC Quality Press   ©1997 by Harris Corporation, Semiconductor Sector

Process Flowchart: Package Assembly

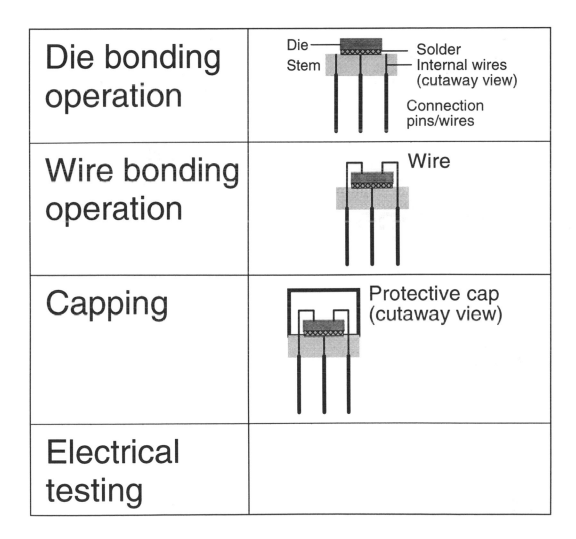

Die bonding operation	Die — Solder Stem — Internal wires (cutaway view) Connection pins/wires
Wire bonding operation	Wire
Capping	Protective cap (cutaway view)
Electrical testing	

Transparency Masters to Accompany *SPC Essentials and Productivity Improvement*
ASQC Quality Press   ©1997 by Harris Corporation, Semiconductor Sector

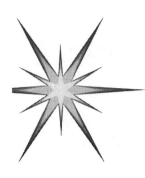

Simple Flowchart

Die bonding

Wire bonding

Capping

Testing

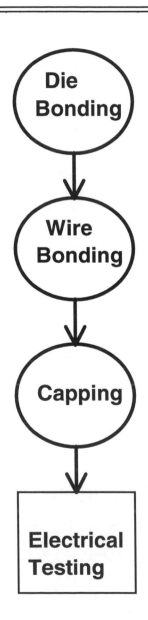

Transparency Masters to Accompany *SPC Essentials and Productivity Improvement*
ASQC Quality Press   ©1997 by Harris Corporation, Semiconductor Sector

Detailed Flowchart (more useful)

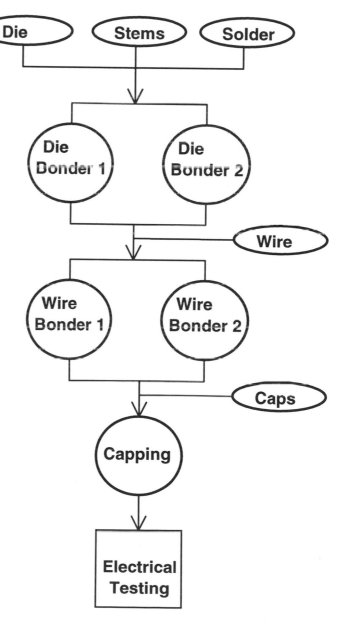

Raw materials

Die   Stems   Solder

Die bonding

Die Bonder 1   Die Bonder 2

Wire

Wire bonding

Wire Bonder 1   Wire Bonder 2

Caps

Capping

Capping

Testing

Electrical Testing

Transparency Masters to Accompany *SPC Essentials and Productivity Improvement*
ASQC Quality Press   ©1997 by Harris Corporation, Semiconductor Sector

Detailed Flowchart, cont.

➤ Suppose that we are getting stoppages at the die bonding operation.

➤ The detailed flowchart prompts these questions.

  ➤ What parts do the stoppages involve? Die? Stems? Solder?

  ➤ Does the problem affect both die bonders or just one?

Critical Processes

- ➤ King Frederick II of Prussia said, "One who tries to protect everything ends up protecting nothing."

  - ➤ Thus, we must concentrate resources and attention on the vital few, not the trivial many.

  - ➤ It is a common mistake in industry to put a statistical control chart on every process.

 # Critical Processes vs. Wallpaper

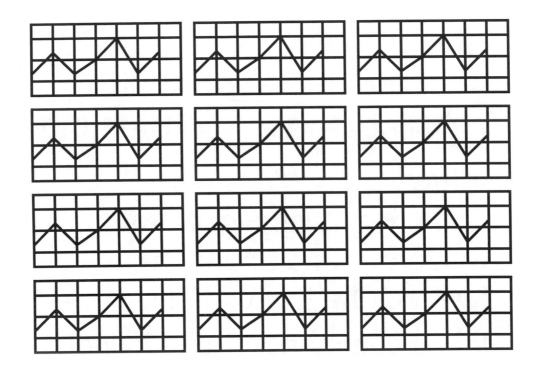

➤ Control charts for every process.

   ➤ No one pays attention to them.

   ➤ They look good on the wall.

Transparency Masters to Accompany *SPC Essentials and Productivity Improvement*
ASQC Quality Press   ©1997 by Harris Corporation, Semiconductor Sector

Critical Processes, cont.

- Control the vital operations that affect productivity and quality.

- A critical process is an operation that can significantly affect

  - Yield

  - Productivity

  - Reliability

  - Performance

Transparency Masters to Accompany *SPC Essentials and Productivity Improvement*
ASQC Quality Press   ©1997 by Harris Corporation, Semiconductor Sector

Cost of Quality

- ➤ Cost of quality (COQ) analysis quantifies the cost of poor quality.
  - ➤ It defines quality problems, or opportunities for improvement, in the language of money.
    - ➤ This is the language of upper management.
    - ➤ It is also useful for quantifying these projects' benefits.
  - ➤ COQ is the cost of poor quality or of avoiding it.
  - ➤ COQ analysis fits in with the process flowchart.
    - ➤ Activities on the flowchart fall into COQ categories.

Cost of Quality, cont.

- ➤ COQ categories
  - ➤ Required (R) activities
    - ➤ These add value to the product.
  - ➤ Appraisal (A) activities
    - ➤ These detect poor quality.
  - ➤ Prevention (P) activities
    - ➤ These prevent poor quality.
  - ➤ Failure (F) costs
    - ➤ These are the costs of rework, scrap, and shipment of bad product.

Transparency Masters to Accompany *SPC Essentials and Productivity Improvement*
ASQC Quality Press   ©1997 by Harris Corporation, Semiconductor Sector

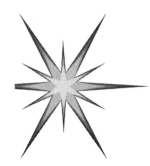

Cost of Quality, cont.

- ➤ Required activities add value to the product.

- ➤ Appraisals detect problems.
  - ➤ Acceptance sampling
  - ➤ Incoming inspection
  - ➤ Final inspection or test

- ➤ Prevention activities prevent problems.
  - ➤ Process control, including SPC
  - ➤ Design for manufacture (DFM), or designing quality into the product
  - ➤ Preventive maintenance such as TPM
  - ➤ Most ISO 9000 activities

Cost of Quality, cont.

- Failure costs
  - Internal failure
    - Rework and scrap
    - 100% inspection (detailing, rectification) of lots that fail acceptance sampling
    - Opportunity costs of rework, scrap, downtime, and so on
  - External failure
    - In the customer's hands
    - Warranty costs, returns
    - Customer dissatisfaction (intangible)

Transparency Masters to Accompany *SPC Essentials and Productivity Improvement*
ASQC Quality Press   ©1997 by Harris Corporation, Semiconductor Sector

Cost of Quality, cont.

➤ The total costs are R + A + P + F.

➤ The cost of quality is A + P + F.

➤ Cost of quality percentage:

$$\frac{A + P + F}{R + A + P + F} \times 100\%$$

➤ In order of increasing desirability

  ➤ Get the disease and suffer from it (external failure, angry customers).

  ➤ Diagnose and cure the disease (appraisal and internal failure).

  ➤ Prevent the disease (prevention).

Transparency Masters to Accompany *SPC Essentials and Productivity Improvement*
ASQC Quality Press   ©1997 by Harris Corporation, Semiconductor Sector

 # Cause-and-Effect Diagram

➤ Also known as the *fishbone diagram* because of its structure, or the *Ishikawa diagram*, after its inventor.

  ➤ A systematic tool for helping a group organize its thoughts

➤ Factors

  ➤ Methods

  ➤ Measurements

  ➤ Manpower

    ➤ Human/personnel factors

  ➤ Materials

  ➤ Machines

  ➤ Environment

    ➤ Or medium, for 6 *M*s instead of 5 *M*s and an *E*

Transparency Masters to Accompany *SPC Essentials and Productivity Improvement*
ASQC Quality Press   ©1997 by Harris Corporation, Semiconductor Sector

Cause-and-Effect Diagram, cont.

- ➤ Methods
  - ➤ Operating instruction or procedure for doing the job
  - ➤ Must be clear and easy to understand
- ➤ Measurements
  - ➤ "If it can't be measured, it can't be controlled."
  - ➤ Gage accuracy, calibration
  - ➤ Gage precision
- ➤ Manpower
  - ➤ Training
  - ➤ Motivation, performance measurements

Transparency Masters to Accompany *SPC Essentials and Productivity Improvement*
ASQC Quality Press   ©1997 by Harris Corporation, Semiconductor Sector

Cause-and-Effect Diagram, cont.

- ➤ Materials
  - ➤ Raw materials and subassemblies
- ➤ Machines (equipment and tools)
  - ➤ Machine or material?
    - ➤ Materials are consumables, or they become part of the product.
    - ➤ Materials appear on the bill of materials (BOM).
    - ➤ The accounting system treats materials as expenses.
    - ➤ Machines are durable items.

Cause-and-Effect Diagram, cont.

- ➤ Environment (Medium)
  - ➤ Temperature
  - ➤ Humidity
- ➤ In the semiconductor/ microelectronics industry, add
  - ➤ Particulates
  - ➤ Noise and vibration
  - ➤ Electrostatic discharge (ESD)
- ➤ In the pharmaceutical and food industries, add
  - ➤ Bacterial contamination

Transparency Masters to Accompany *SPC Essentials and Productivity Improvement*
ASQC Quality Press   ©1997 by Harris Corporation, Semiconductor Sector

Die Bonder Example

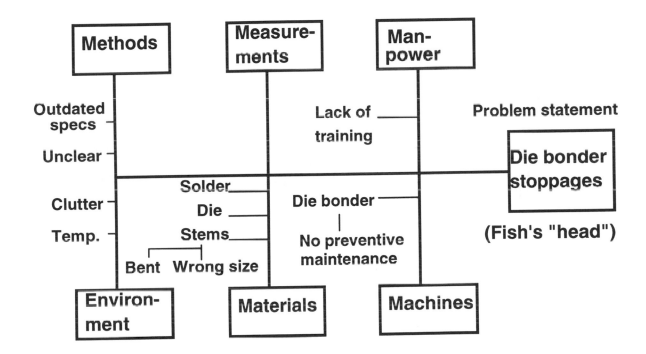

Methods	Measure-ments	Man-power

Outdated specs

Unclear

Lack of training

Problem statement

Die bonder stoppages

Solder

Die

Die bonder

Clutter

Temp.

Stems

No preventive maintenance

(Fish's "head")

Bent   Wrong size

Environ-ment	Materials	Machines

➤ The diagram's branching feature prompts the group to look for underlying causes.

Transparency Masters to Accompany *SPC Essentials and Productivity Improvement*
ASQC Quality Press   ©1997 by Harris Corporation, Semiconductor Sector

Die Bonder Example: Branching

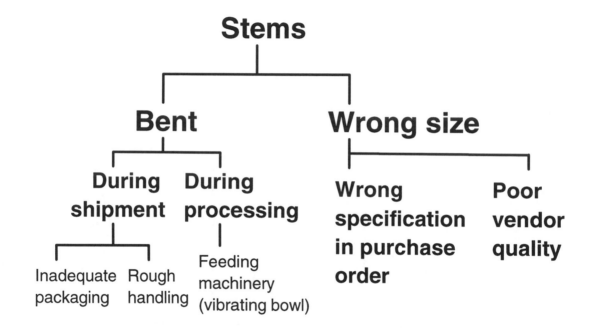

Stems

Bent

During shipment

Inadequate packaging   Rough handling

During processing

Feeding machinery (vibrating bowl)

Wrong size

Wrong specification in purchase order

Poor vendor quality

➤ The entries for stems prompt the team to think about why the stems could jam the machine.

   ➤ The team should continue to follow each branch until it runs out of ideas.

Case Study: Wafer Breakage

- ➤ Semiconductor manufacturing involves coating silicon wafers with photosensitive films.

- ➤ Silicon wafers are brittle.

- ➤ A Harris Semiconductor manufacturing team was experiencing wafer breakage during this process.

- ➤ The team set up wafer checkpoints at five process steps.

    - ➤ The team used a check sheet to collect data for one month.

    - ➤ The team made a Pareto chart of the data.

Transparency Masters to Accompany *SPC Essentials and Productivity Improvement*
ASQC Quality Press   ©1997 by Harris Corporation, Semiconductor Sector

Wafer Breakage, cont.

Defect	Abbreviation	Count
Wafers broken at coating step 1	Broken, coater 1	29
Wafers broken at coating step 2	Broken, coater 2	4
Operator error	Operator	5
Wafers broken during deposition	Broken dep	6
Miscellaneous	Misc	1

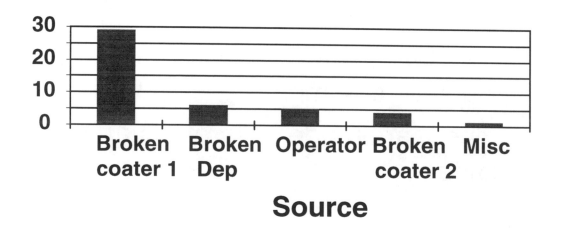

Source

Wafer Breakage, cont.

- ➤ The team developed a cause-and-effect diagram.
  - ➤ The operator has to pick up the wafer with tweezers.
  - ➤ Wafers have become larger and heavier over the years.
  - ➤ The spin coater spins the wafer at 5000 rpm.
  - ➤ The wafer goes on a hot plate to drive off the solvent.

Wafer Breakage, cont.

➤ Here is the cause-and-effect diagram for wafer breakage.

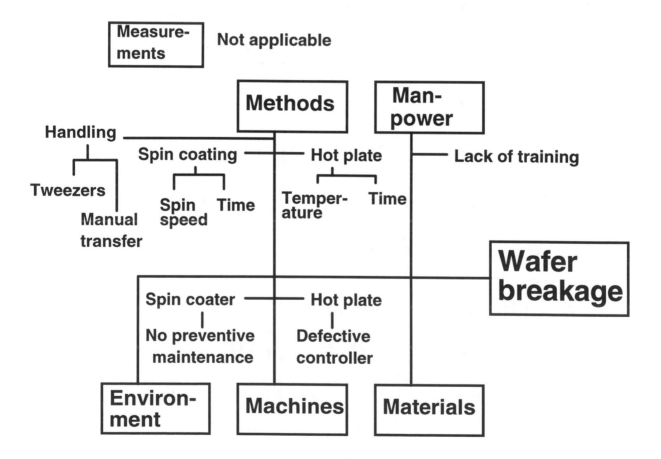

Transparency Masters to Accompany *SPC Essentials and Productivity Improvement*
ASQC Quality Press   ©1997 by Harris Corporation, Semiconductor Sector

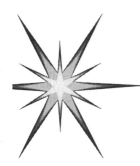

Wafer Breakage, cont.

- ➤ The team tried the following actions.
  - ➤ Purchase a wafer transfer machine to replace manual handling with tweezers.
    - ➤ When wafers were smaller, a very light grip was adequate to hold them. The team questioned whether tweezers are appropriate for handling large wafers.
  - ➤ Replace tweezers with vacuum wands in other operations.
    - ➤ A vacuum wand is a flat tool that holds the back of the wafer by suction.

Transparency Masters to Accompany *SPC Essentials and Productivity Improvement*
ASQC Quality Press   ©1997 by Harris Corporation, Semiconductor Sector

Wafer Breakage, cont.

> Actions, cont.

> > Change the spin speed from 5000 rpm to 2500 rpm, and raise the spin time to get the same coating thickness.

> > Lower the temperature to 80°C.

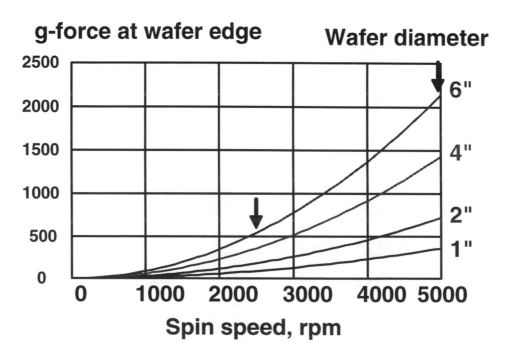

g-force at wafer edge   **Wafer diameter**

Spin speed, rpm

Transparency Masters to Accompany *SPC Essentials and Productivity Improvement*
ASQC Quality Press   ©1997 by Harris Corporation, Semiconductor Sector

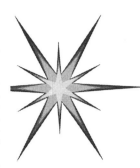

Wafer Breakage, cont.

➤ Here is the Pareto chart after the changes went into effect.

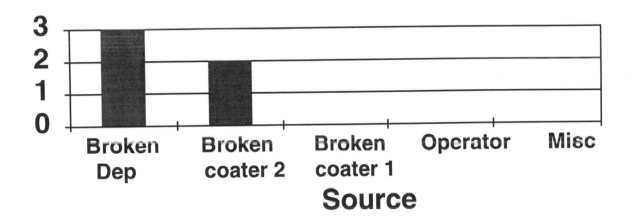

➤ Why is operator error now zero?

➤ Was it the operator or the tool (tweezers)?

Transparency Masters to Accompany *SPC Essentials and Productivity Improvement*
ASQC Quality Press   ©1997 by Harris Corporation, Semiconductor Sector

Team-Oriented Problem Solving

➤ Ford Motor Company's TOPS-8D is a systematic approach for solving problems.

  ➤ It contains all the elements of the plan-do-check-act (PDCA) cycle.

  ➤ It is particularly suitable for self-directed work teams.

➤ 1. Use a team approach.

  ➤ The team should have a champion. This person should have ownership of the problem.

    ➤ Ownership = Resources and authority to implement the team's decisions.

    ➤ The champion is not necessarily the team leader.

Transparency Masters to Accompany *SPC Essentials and Productivity Improvement*
ASQC Quality Press   ©1997 by Harris Corporation, Semiconductor Sector

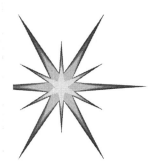

2. Working Definition

- Define the problem. Compare what should have happened to what actually happened.
  - Define and identify the problem clearly.
    - Make sure that everyone is trying to solve the same problem!
- Ask what, where, when, and how big.
  - What is the problem?
  - Where is it seen or detected?
  - When is it seen?
  - How big is it? Quantify the problem's severity.

Transparency Masters to Accompany *SPC Essentials and Productivity Improvement*
ASQC Quality Press   ©1997 by Harris Corporation, Semiconductor Sector

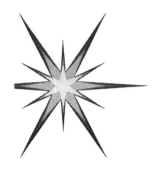

2. Working Definition, cont.

➤ An "is/is not" description is helpful.

  ➤ Knowing what the problem isn't, as well as what it is, helps us locate its source.

  ➤ Combine the "is/is not" question with what, where, when, and how big?

➤ Where does detection or identification of the defect happen?

  ➤ Is it detectable at its source?

  ➤ The process flow diagram is useful here.

Transparency Masters to Accompany *SPC Essentials and Productivity Improvement*
ASQC Quality Press   ©1997 by Harris Corporation, Semiconductor Sector

2. Working Definition
3. Containment

- What is the defect's source?
  - Which process step generates it?
  - Is the defect traceable to a particular workstation?
    - The detailed process flow diagram shows material flows and multiple workstations.

- Contain the problem.
  - Containment is similar to quarantine. Find and segregate nonconforming products to prevent their shipment to customers.
  - Shut down manufacturing equipment that is making bad product.
  - Make sure the containment action is effective.

Transparency Masters to Accompany *SPC Essentials and Productivity Improvement*
ASQC Quality Press   ©1997 by Harris Corporation, Semiconductor Sector

4. Identify Root Cause
5. Permanent Correction

- ➤ This is like diagnosing the disease.
- ➤ Tools for doing this include the check sheet, Pareto chart, and cause-and-effect diagram.
- ➤ Design of experiments (DOE or DOX) is a quantitative tool for investigating a problem.
  - ➤ It requires the aid of someone with statistical training.
- ➤ Select and carry out a permanent correction for the root cause.
  - ➤ This is like curing the disease.
  - ➤ Make sure that the action is effective.

5, 6, and 7. Permanent Correction and 8. Recognition

➤ Prevent the problem from coming back.

  ➤ Hold the gains.

    ➤ Implement controls (statistical or otherwise) to make sure the improvement is permanent.

➤ Recognize the team's accomplishment.

  ➤ Each company has its own procedures for doing this.

  ➤ Harris Semiconductor uses bulletin boards and monthly communication meetings.

  ➤ Recognition educates and encourages other teams.

Transparency Masters to Accompany *SPC Essentials and Productivity Improvement*
ASQC Quality Press   ©1997 by Harris Corporation, Semiconductor Sector

"Is/Is Not" Analysis

➤ Knowing what a problem isn't, as well as what it is, helps isolate its source.

➤ We will use the die bonder stoppage problem as an example.

  ➤ Is/is not analysis works well with process flowcharts and cause-and-effect diagrams.

"Is/Is Not," cont.

	Is	Is not
What	Die bonder, stoppage	Wire bonder or capper stoppage; a soldering problem
Where	**Both wire bonding stations** Stem handling mechanism	**Only one station** Die handling mechanism
When	Intermittent	Consistent
How big		

Transparency Masters to Accompany *SPC Essentials and Productivity Improvement*
ASQC Quality Press   ©1997 by Harris Corporation, Semiconductor Sector

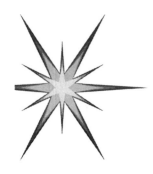

"Is/Is Not," cont.

- ➤ The entries for "where" help us rule out a die bonder problem as a root cause.
  - ➤ It's unlikely that both stations would have the same problem at the same time.
    - ➤ The process flowchart prompted the question, "Both stations or just one?"
  - ➤ The problem does not affect the wire bonder or the capper.
  - ➤ The caps, wire, and solder are not possible root causes.
- ➤ This analysis points to either the die or the stem.

Transparency Masters to Accompany *SPC Essentials and Productivity Improvement*
ASQC Quality Press   ©1997 by Harris Corporation, Semiconductor Sector

"Is/Is Not," cont.

➤ Now consider the following. What is the likely problem?

	Is	Is not
What	Die bonder; stoppage	Wire bonder or capper stoppage; a soldering problem
Where	**Only one station** Stem handling mechanism	**Both wire bonding stations** Die handling mechanism
When	Intermittent	Consistent
How Big		

Transparency Masters to Accompany *SPC Essentials and Productivity Improvement*
ASQC Quality Press   ©1997 by Harris Corporation, Semiconductor Sector

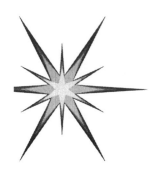

"Is/Is Not," cont.

➤ Tabulate the theories against the is/is not entries.

   ➤ **+** Explains both the "is" and "is not."

      ➤ To be plausible, a root cause should have mostly (preferably all) plus (+) signs.

   ➤ **—** Is inconsistent with the "is" or the "is not" description.

   ➤ **?** Need more information

Transparency Masters to Accompany *SPC Essentials and Productivity Improvement*
ASQC Quality Press ©1997 by Harris Corporation, Semiconductor Sector

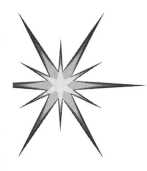

"Is/Is Not," cont.

	Is	Is not	Possible root cause		
			Die bonder	Stems	Die
What	Die bonder, stoppage	Wire bonder or capper stoppage	+	+	+
Where	Both wire bonding stations	Only one station	-	+	+
	Stem handling mechanism	Die handling mechanism	+	+	-
When	Intermittent	Consistent	+	+	+
How big					

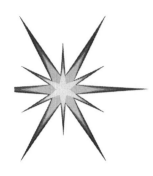

"Is/Is Not," cont.

➤ "Problem with the stems" is the only root cause in the table that has no "–" symbols.

➤ It is the only explanation that is consistent with all the information.

➤ This does not prove that stems are the problem source, but it tells us where to look.

Transparency Masters to Accompany *SPC Essentials and Productivity Improvement*
ASQC Quality Press   ©1997 by Harris Corporation, Semiconductor Sector

ISO 9000

➤ Murphy's Law says, "Anything that can go wrong, will."

➤ ISO 9000 is a system for making sure that everything will go right.

    ➤ ISO 9000 is a set of international standards for quality management systems.

➤ The automotive QS-9000 standards extend ISO 9000.

    ➤ An organization that qualifies for QS-9000 certification should also meet ISO 9000 standards.

    ➤ Meeting ISO 9000 standards, however, is not sufficient for QS-9000 certification.

Transparency Masters to Accompany *SPC Essentials and Productivity Improvement*
ASQC Quality Press   ©1997 by Harris Corporation, Semiconductor Sector

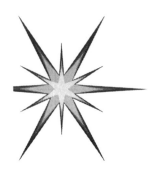

ISO 9000, cont.

- ➤ ISO 9000 systematically guides an assessment of our quality management system and process controls.

  - ➤ It makes us ask, "What can go wrong, and how can we change the system to prevent it from doing so?"

Document Control

- ➤ Document control assures that everyone does the job the same way.
  - ➤ Is everyone using the same work instructions?
- ➤ Elements of document control include the following:
  - ➤ Assure that work instructions and process recipes are up to date.
  - ➤ Notify people when documents change.
    - ➤ Maintain a document review log.
    - ➤ To avoid confusion, make sure that the review log shows the revision number.
  - ➤ Remove obsolete documents from the factory.

Transparency Masters to Accompany *SPC Essentials and Productivity Improvement*
ASQC Quality Press   ©1997 by Harris Corporation, Semiconductor Sector

Product Identification/ Traceability

- ➤ If our process is making defective product, we need to know from where it is coming.

  - ➤ Do all the bad parts come from one workstation or one material lot?

- ➤ Traceability is also a requirement for effective SPC.

Process Control

- ➤ Process control includes
  - ➤ Control of the working environment
    - ➤ Temperature and humidity
    - ➤ Noise and vibration
    - ➤ Particulates
  - ➤ Control of the manufacturing process
    - ➤ SPC
    - ➤ Feedback process control
  - ➤ Preventive maintenance
    - ➤ Total productive maintenance

# Inspection, Testing, Calibration

- ➤ Assure that products that require testing or inspection receive them.
    - ➤ Outgoing products
    - ➤ Incoming materials
- ➤ Gages (measurement equipment, instruments) need periodic calibration.
    - ➤ Reconditioning = calibration
    - ➤ Calibration schedule

Transparency Masters to Accompany *SPC Essentials and Productivity Improvement*
ASQC Quality Press   ©1997 by Harris Corporation, Semiconductor Sector

Calibration, cont.

➤ Calibration is like the chain of custody for evidence.

Vehicle on police evidence lot

Status when checked:

 **Day 1; O.K.**

 **Day 4; O.K.**

Evidence collected after this is unreliable.

 **Day 10; discovery of burglary/ vandalism**

Transparency Masters to Accompany *SPC Essentials and Productivity Improvement*
ASQC Quality Press   ©1997 by Harris Corporation, Semiconductor Sector

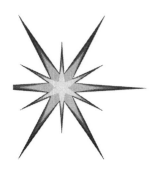

Calibration, cont.

➤ A gage that is out of calibration is a major quality exposure.

Shipments

March 1; in calibration | O.K.

| ? |

| ? |

| ? |

April 1; out of calibration | ? |

? = Surprise package for customer
(probably an unpleasant surprise)

Transparency Masters to Accompany *SPC Essentials and Productivity Improvement*
ASQC Quality Press   ©1997 by Harris Corporation, Semiconductor Sector

Calibration, cont.

- ➤ Calibration stickers show when a gage needs calibration.
  - ➤ Date of last calibration
  - ➤ Initials of person who did it
  - ➤ Date when the next calibration is due
- ➤ Gages that do not require calibration need stickers saying, "No calibration required."
- ➤ Uncalibrated gages that are not intended for product, but could be used inadvertently, should not be present in the manufacturing area.

Control of Nonconforming Product

➤ C.S. Forester example from *Horatio Hornblower* stories

(1) Barrel contents: You don't want to know.

(2) Will the supplier try to resell them?

(3) Mark each barrel "Condemned"

Nonconforming Product, cont.

- ➤ Identify and segregate nonconforming items to prevent their accidental use.

  - ➤ A bright orange **REJECTED** sticker can identify nonconforming materials or parts.

  - ➤ A reject rack, or lockable reject cage, can hold nonconforming product.

Nonconforming Product, cont.

➤ Example: An electrical tester sorts parts.

   ➤ The arm's idle position is over the reject bin.

   ➤ The tester counts the nonconforming pieces.

      ➤ It must match the reject bin count.

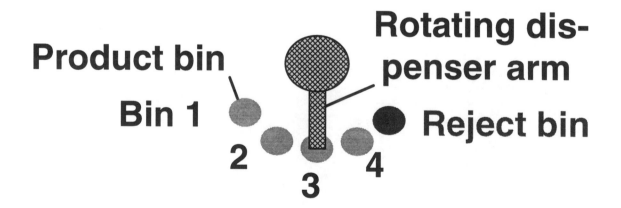

Rotating dis-penser arm

Product bin

Bin 1

2

3

4

Reject bin

# Storage, Packaging, Handling, Delivery

Exposure to salt water
(Red Storm Rising)

 ESD

 Transistor package

Tilt indicator (permanent change)

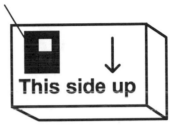

Use by 12/15/95

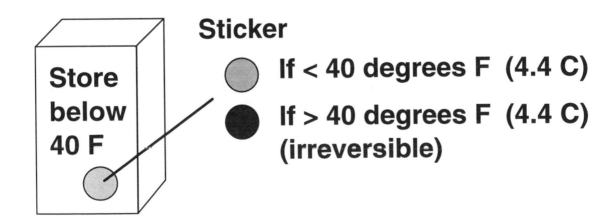

Sticker

Store below 40 F

If < 40 degrees F (4.4 C)

If > 40 degrees F (4.4 C)
(irreversible)

Transparency Masters to Accompany *SPC Essentials and Productivity Improvement*
ASQC Quality Press   ©1997 by Harris Corporation, Semiconductor Sector

Storage, Packaging, Handling, Delivery, cont.

- ➤ Is the product perishable?
- ➤ Does it have a shelf life?
  - ➤ Expiration dates on food items in supermarkets
- ➤ Is it sensitive to heat?
  - ➤ Temperature-sensitive stickers for packages
- ➤ Is it sensitive to shock?
  - ➤ Tipping indicators show whether "This Side Up" has always been up.

Transparency Masters to Accompany *SPC Essentials and Productivity Improvement*
ASQC Quality Press   ©1997 by Harris Corporation, Semiconductor Sector

Chapter 2, Problem 1

➤ The following defect counts occurred at a preprobe silicon wafer inspection. Make a Pareto chart of the data.

Defect	Abbreviation	Count
Blisters	BLIS	4
Discoloration	DISC	52
Oxide holes	OX	14
Damage	DAM	163

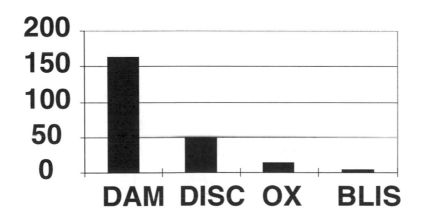

Transparency Masters to Accompany *SPC Essentials and Productivity Improvement*
ASQC Quality Press   ©1997 by Harris Corporation, Semiconductor Sector

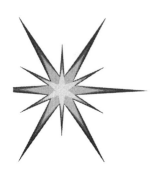

Chapter 2, Problem 2

➤ Make a Pareto chart of the following data.

Defect	Abbreviation	Count
Blisters	BLIS	31
Discoloration	DISC	2
Oxide holes	OX	20
Damage	DAM	5
Miscellaneous	MISC	2

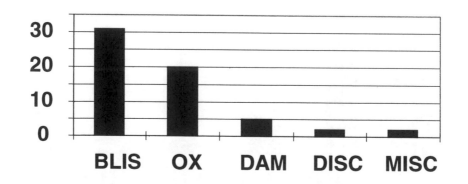

Transparency Masters to Accompany *SPC Essentials and Productivity Improvement*
ASQC Quality Press   ©1997 by Harris Corporation, Semiconductor Sector

Chapter 2, Problem 3

➤ Make a Pareto chart of the following data.

Operation	Abbreviation	Scratches
Silicon nitride etching	NitE	97
Aluminum etching and stripping	Etch	94
Wafer probing (electrical test)	Prob	34
Backside metal deposition	BMet	23
Silicon nitride deposition	NitD	22
Ion implantation	Imp	11
Back grinding	Grind	2
Wafer firing	Fire	0

Chapter 2, Problem 3, cont.

Operation	Abbreviation	Scratches
Silicon nitride etching	NitE	97
Aluminum etching and stripping	Etch	94
Wafer probing (electrical test)	Prob	34
Backside metal deposition	BMet	23
Silicon nitride deposition	NitD	22
Ion implantation	Imp	11
Back grinding	Grind	2
Wafer firing	Fire	0

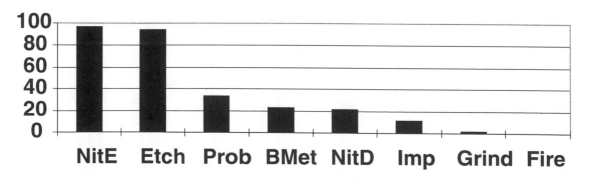

Transparency Masters to Accompany *SPC Essentials and Productivity Improvement*
ASQC Quality Press   ©1997 by Harris Corporation, Semiconductor Sector

Chapter 2, Problem 4

- A factory has three shifts and works 24 hours a day.

  - The machine is available 24 hours a day.

  - The machine operates, on average, 18 hours a day.

  - The machine can handle 50 pieces. The usual load is 40 pieces.

  - In each load, 2 pieces out of the 40 are, on average, reworks.

Transparency Masters to Accompany *SPC Essentials and Productivity Improvement*
ASQC Quality Press   ©1997 by Harris Corporation, Semiconductor Sector

Chapter 2, Problem 4, cont.

➤ Compute the overall equipment effectiveness (OEE) for the machine. The machine is not at a constraint operation.

    ➤ Are the partial loads or idle time problems?

    ➤ What about the rework?

Chapter 2, Problem 4, cont.

Availability	100%, since the tool is always available.
Operating efficiency	$\dfrac{18 \text{ hours / day (operating)}}{24 \text{ hours / day (availability)}} \times 100\% = 75\%$
Rate efficiency	$\dfrac{40 \text{ pieces / load (actual)}}{50 \text{ pieces / load (capacity)}} \times 100\% = 80\%$
Rate of quality	$\dfrac{38 \text{ good pieces / load}}{40 \text{ pieces / load}} \times 100\% = 95\%$
OEE	$75\% \times 80\% \times 95\% = 57\%$, or $$\text{OEE} = \frac{\text{Operating time}}{\text{Total time}} \times$$ $$\frac{\text{Good pieces}}{\text{Theoretical output}} \times 100\%$$ $$\frac{18 \text{ hours / day}}{24 \text{ hours / day}} \times \frac{38 \text{ good pieces}}{50 \text{ pieces / load}}$$ $$\times 100\% = 57\%$$

Transparency Masters to Accompany *SPC Essentials and Productivity Improvement*
ASQC Quality Press   ©1997 by Harris Corporation, Semiconductor Sector

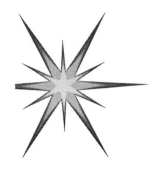

Chapter 2, Problem 4, cont.

➤ Since this operation is not a constraint, don't worry about the idle time (operating efficiency) or partial loads (rate efficiency).

➤ Rework, however, is never desirable. The rate of quality goal should be 100%.

Chapter 2, Problem 5

➤ Classify the following activities or events as required, appraisal, prevention, internal failure, or external failure (cost of quality analysis).

➤ a. An acceptance sampling plan examines a sample of 200 pieces. If one or fewer are bad, the lot passes. If not, it fails.

➤ b. A lot fails the above acceptance sampling inspection, so it receives 100% inspection to remove all the bad units.

➤ c. An assembly operation puts two subassemblies together.

Transparency Masters to Accompany *SPC Essentials and Productivity Improvement*
ASQC Quality Press   ©1997 by Harris Corporation, Semiconductor Sector

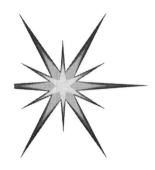

Chapter 2, Problem 5 cont.

- ➤ d. An SPC chart shows when the process average has shifted.

- ➤ e. An office chair breaks. The customer identifies the part that needs replacement, but the manufacturer's customer service department does not respond to phone calls or letters. The customer must replace the chair, buys a competitor's product, and tells the original supplier he will never buy anything from it again.

- ➤ f. A transistor assembly process solders the transistor to a stem, and places a cap over the transistor.

Chapter 2,
Problem 5 cont.

- ➤ g. An electrical tester checks the finished units from item F and rejects the bad ones. Also, classify the bad pieces.

- ➤ h. A product's designers work closely with manufacturing to make sure the product is easy to manufacture. This is design for manufacture (DFM).

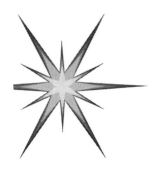

Chapter 2, Problem 6

➤ Are the following actions or procedures acceptable under ISO 9000? Why or why not? What should be changed?

➤ a. A process' operating instruction calls for a 30-minute process time. A manager or engineer verbally instructs a manufacturing shift to set the time to 25 minutes.

➤ b. A process' operating instruction calls for a 30-minute process time. A manager or engineer makes a handwritten change in the operating instructions to set the time to 25 minutes.

Chapter 2,
Problem 6 cont.

➤ c. Operators place bright orange REJECTED stickers on nonconforming pieces and place them on a dedicated shelf.

➤ d. A factory work area has three shifts, and 12 people do a particular job. An engineer removes the old set of work instructions and replaces them with new ones. The new instructions have the necessary approval from the manufacturing manager. What else, if anything, should have happened?

Chapter 2,
Problem 6 cont.

➤ e. Operators normally use a calibrated electronic micrometer to measure parts. A manual micrometer is sitting on a table in the work area. There is also a window thermometer, but the work area is not subject to temperature and humidity controls.

➤ f. Subassemblies receive bar codes to identify them. The operator scans the bar code at each operation, and a computer logs the subassembly identification and workstation identification. When an operator measures a piece, he or she logs the measurement and scans the bar code.

Transparency Masters to Accompany *SPC Essentials and Productivity Improvement*
ASQC Quality Press   ©1997 by Harris Corporation, Semiconductor Sector

Statistical Process Control

Variation and Accuracy

Transparency Masters to Accompany *SPC Essentials and Productivity Improvement*
ASQC Quality Press   ©1997 by Harris Corporation, Semiconductor Sector

Variation and Accuracy

➤ Two factors define a manufacturing process' ability to meet specifications: variation and accuracy.

  ➤ Precision is the opposite of variation.

➤ A specification is like a target.

  ➤ Inside the target is good.

  ➤ Outside the target is bad.

Ye Musket (high variation tool)

➤ Accurate: The aiming point is the bull's-eye.

➤ Not precise: There is a lot of variation.

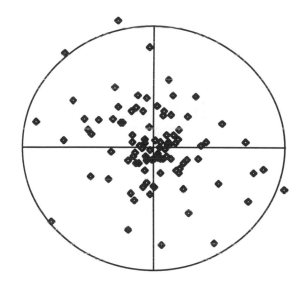

Transparency Masters to Accompany *SPC Essentials and Productivity Improvement*
ASQC Quality Press   ©1997 by Harris Corporation, Semiconductor Sector

Can We Improve This Process?

- ➤ Which action(s) would help?
  - ➤ Adjust the musket's sights (adjust the process).
  - ➤ Hire a better marksman: maybe Daniel Boone or Annie Oakley.
  - ➤ Replace the musket with a rifle.
- ➤ The musket is not capable of better performance.
  - ➤ Adjusting it will only make it worse. (It's aimed at the bull's-eye.)
  - ➤ Don't blame the operator.
  - ➤ To improve this process, get a rifle (low variation/high precision tool).

Transparency Masters to Accompany *SPC Essentials and Productivity Improvement*
ASQC Quality Press   ©1997 by Harris Corporation, Semiconductor Sector

The Rifle (low variation tool)

➤ Accurate:
Aimed at the
target
center.

➤ Capable:
Precise,
low variation

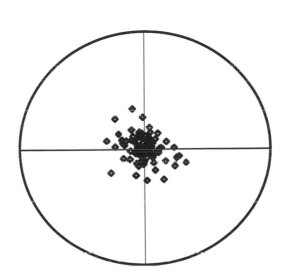

Transparency Masters to Accompany *SPC Essentials and Productivity Improvement*
ASQC Quality Press   ©1997 by Harris Corporation, Semiconductor Sector

# Process Capability

➤ A process' *capability index* measures its ability to meet specifications (hit the target).

    ➤ Capable processes are precise (have low variation).

    ➤ The process capability index uses the ratio of the specification width (target size) to the tool variation.

$$\text{Process Capability Index} = \frac{\text{Specification Width}}{\text{Variation}}$$

Process Capability Indices

Process capability index	Process status	Bad pieces per million	Similar to a
Less than 1.00	Not capable	2700 or more	Musket
1.00 to 1.33	Marginal	2700 to 63.3	
1.33 to 2.00	Capable	63.3 to 0.002 (2 per billion)	Rifle
2.00 or better	Very capable	0.002 or less	Olympic match rifle

Transparency Masters to Accompany *SPC Essentials and Productivity Improvement*
ASQC Quality Press   ©1997 by Harris Corporation, Semiconductor Sector

High Variation Process vs. Wide Specification

➤ English musket: rapid fire, high variation

➤ Enemy in shoulder-to-shoulder formation: big target (wide spec)

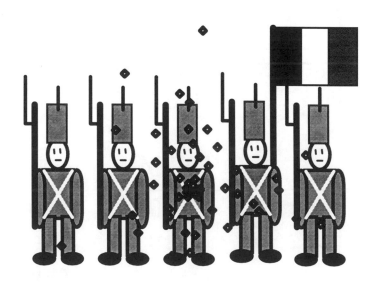

Transparency Masters to Accompany *SPC Essentials and Productivity Improvement*
ASQC Quality Press   ©1997 by Harris Corporation, Semiconductor Sector

High Variation Process vs. Tight Specification

> Simulation: 50 musket shots in one minute vs. small target/tight specification, about 6 hits

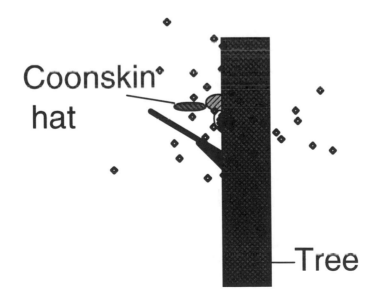

Coonskin° hat

Tree

Transparency Masters to Accompany *SPC Essentials and Productivity Improvement*
ASQC Quality Press   ©1997 by Harris Corporation, Semiconductor Sector

Low Variation Process vs. Wide Specification

- ➤ Simulation: 10 rifle shots in one minute, against a wide specification (big target)

- ➤ 9 to 10 hits

Transparency Masters to Accompany *SPC Essentials and Productivity Improvement*
ASQC Quality Press   ©1997 by Harris Corporation, Semiconductor Sector

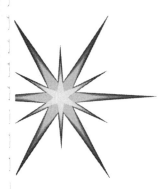

Tightening Specifications

➤ New industrial products often have tight specifications.

    ➤ This is especially true in the microelectronics industry.

➤ Tight specifications demand processes with high precision.

Tightening Specifications, cont.

➤ Example: Semiconductor wiring specifications

  ➤ Left edge of wire goes on heavy black line. Right edge must be between lines at right.

  ➤ 6 micron wire; early 1980s (shoulder-to-shoulder formation).

  ➤ 0.5 micron; mid-1990s (American revolutionary in coonskin cap behind tree).

  [5.70, 6.30] micron specification

  [0.475, 0.525] micron specification

Quality Versus Quantity?

➤ Is there a trade-off between quality and quantity?

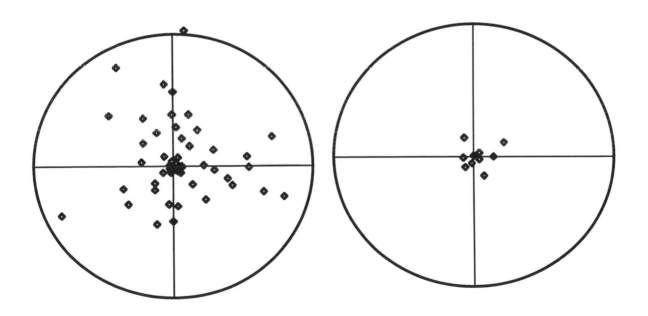

Transparency Masters to Accompany *SPC Essentials and Productivity Improvement*
ASQC Quality Press   ©1997 by Harris Corporation, Semiconductor Sector

Quality and Quantity

➤ The goal of a manufacturing process is to deliver quality and quantity.

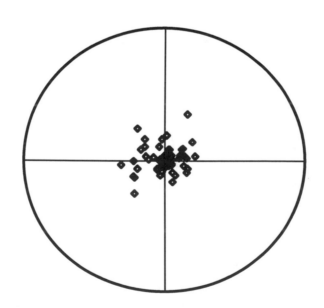

Accuracy

➤ Accuracy, or aiming at the center of the specification (target), is another factor that affects quality.

➤ The *nominal* measurement (bull's-eye) is halfway between the upper and lower specification limits.

Process Shift

➤ The rifle is precise (low variation) but not accurate (not aimed at the bull's-eye).

➤ It is capable but out of control.

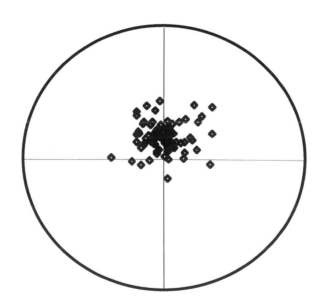

Transparency Masters to Accompany *SPC Essentials and Productivity Improvement*
ASQC Quality Press   ©1997 by Harris Corporation, Semiconductor Sector

Capability and Control

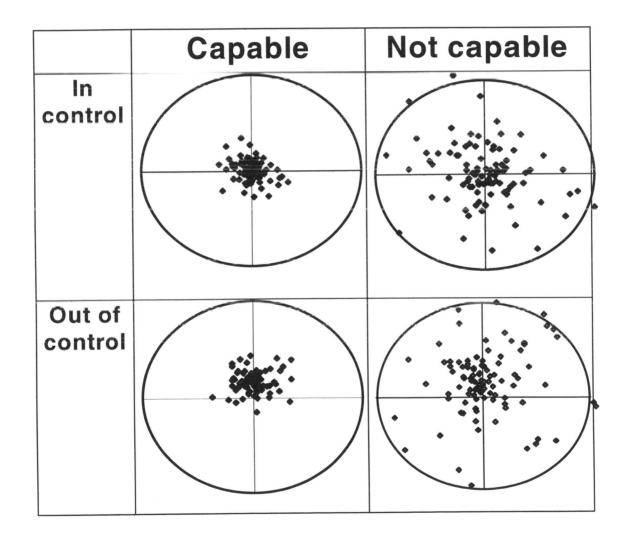

	Capable	Not capable
In control		
Out of control		

Transparency Masters to Accompany *SPC Essentials and Productivity Improvement*
ASQC Quality Press   ©1997 by Harris Corporation, Semiconductor Sector

Process Adjustments

- A process that is out of control needs adjustment.

- To adjust a rifle, the shooter moves the back sight.

- Adjusting a manufacturing process brings it back on target. For example:

    - Change the process time.

    - Change a gas flow.

    - Replenish a chemical.

- SPC tells us when to adjust the process.

Transparency Masters to Accompany *SPC Essentials and Productivity Improvement*
ASQC Quality Press   ©1997 by Harris Corporation, Semiconductor Sector

Samples for SPC

➤ A target shooter would use a group of three to five shots to decide whether to adjust the sights.

  ➤ Here, the sight is too high.

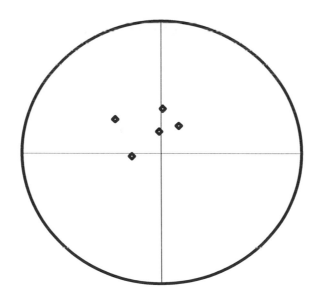

Transparency Masters to Accompany *SPC Essentials and Productivity Improvement*
ASQC Quality Press   ©1997 by Harris Corporation, Semiconductor Sector

# Samples for SPC

➤ In SPC, we take a sample of process measurements.

➤ The sample plays the same role as the group of rifle shots. It tells us whether to change the process or just let it run.

# Overadjustment or Tampering

➤ Don't adjust the process when it's running properly!

   ➤ Instead of adjusting the rifle sights after three to five shots, why not adjust it after every shot? If the shot is 1" high and 1" right, move the aiming point 1" left and down.

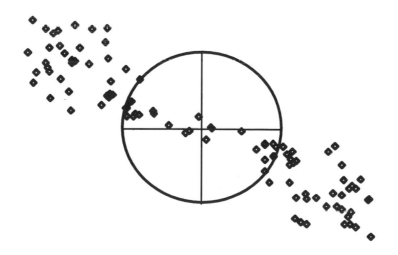

Overadjustment or Tampering, cont.

➤ What is an *iatrogenic disease*? *Iatros* = Greek for "doctor;" *genic* = "caused by."

➤ Prescribing medicine for a patient who isn't sick can cause an iatrogenic disease.

  ➤ Adjusting a process that is running properly is like prescribing unnecessary medication.

  ➤ At best, it wastes time.

  ➤ It can add variation to the process, and make its performance worse.

Transparency Masters to Accompany *SPC Essentials and Productivity Improvement*
ASQC Quality Press   ©1997 by Harris Corporation, Semiconductor Sector

Role of SPC

➤ Diagnostic tests tell the doctor whether the patient is sick and needs medicine.

➤ SPC tells us when to adjust the process and when to leave it alone.

Common and Assignable Causes

➤ Manufacturing problems (rework, scrap, and defects) come from two sources.

  1. *Common causes* or *random variation*

    ➤ These are inherent in the process.

    ➤ The only way to reduce or get rid of random variation is to improve the process.

  2. *Assignable causes* or *special causes*

    ➤ These are problems with the process.

Transparency Masters to Accompany *SPC Essentials and Productivity Improvement*
ASQC Quality Press   ©1997 by Harris Corporation, Semiconductor Sector

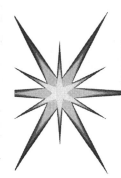

Improvement Versus Correction

Examples of improvement	Examples of correction (Fixing)
Evolution of ground transportation: • Horse and buggy • Early automobile (Model T) • Ford Taurus, Chevrolet Caprice, Chrysler Concorde • Future: George Jetson's car?	• Call veterinarian to cure sick horse. • Take car to garage to fix leaking radiator.
Evolution of electronic technology: • Vacuum tube • Transistor • Integrated circuit	Replace burned-out vacuum tube.
Musket ==> Rifle	Adjust misaligned rifle sights.

Transparency Masters to Accompany *SPC Essentials and Productivity Improvement*
ASQC Quality Press   ©1997 by Harris Corporation, Semiconductor Sector

Conclusion: Common/ Assignable Causes

➤ To reduce or get rid of common causes, we must improve the process.

➤ To remove an assignable cause, we must fix the process.

➤ We should always look for ways to improve the process. SPC tells us whether we need to fix the process.

Transparency Masters to Accompany *SPC Essentials and Productivity Improvement*
ASQC Quality Press   ©1997 by Harris Corporation, Semiconductor Sector

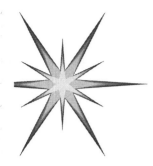

The Normal Distribution

➤ Measurements from most manufacturing processes follow the normal (bell curve) distribution.

➤ There is a relationship between the normal distribution and the shot patterns on the targets.

# Normal Distribution: Example

➤ A semiconductor manufacturing operation deposits a 1000 angstrom (Å) film of silicon dioxide on a silicon wafer.

➤ Because of variation, we know that not all the wafers will get exactly 1000 Å of silicon dioxide.

  ➤ If we process 500 wafers, how many will have a film of 970 Å to 980 Å? How many will have 1020 Å or more?

  ➤ If the specification is [950, 1050], how many will be out of specification?

Transparency Masters to Accompany *SPC Essentials and Productivity Improvement*
ASQC Quality Press   ©1997 by Harris Corporation, Semiconductor Sector

Parameters of the Normal Distribution

- ➤ Mean
  - ➤ Center of gravity or aiming point
  - ➤ Similar to the average
    - ➤ Mean is for the process.
    - ➤ Average is for a sample.
- ➤ Variation
  - ➤ Amount of spread in the process

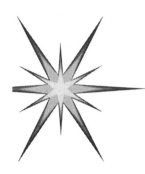

Histogram: High Variation Process

➤ **Mean = 1000 Å, σ = 20**

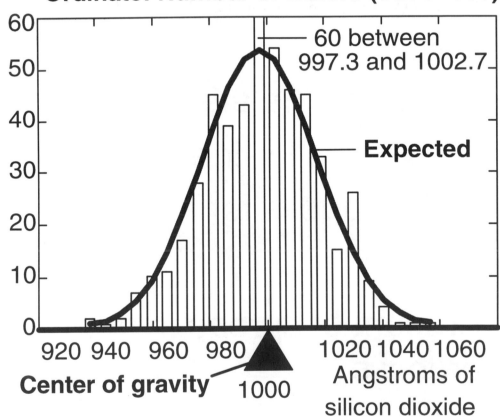

Ordinate: Number of wafers (out of 500)

60 between
997.3 and 1002.7

Expected

920 940 960 980 1020 1040 1060

Center of gravity 1000 Angstroms of
silicon dioxide

Silicon dioxide layer
Silicon wafer (side view)

Transparency Masters to Accompany *SPC Essentials and Productivity Improvement*
ASQC Quality Press   ©1997 by Harris Corporation, Semiconductor Sector

Histograms and Targets

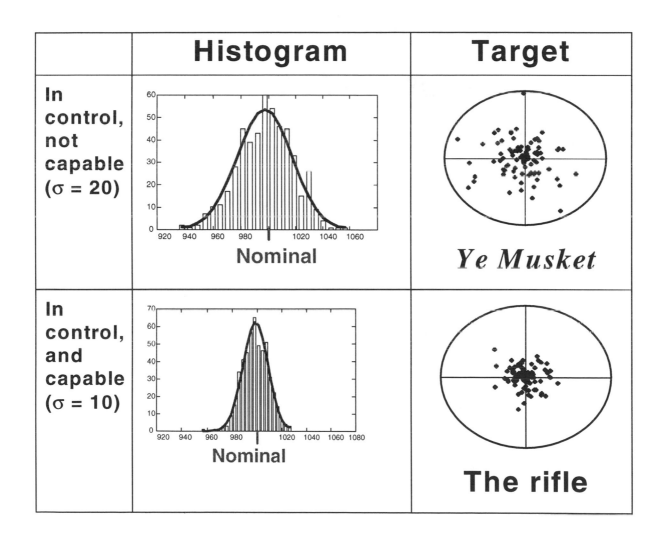

	Histogram	Target
In control, not capable ($\sigma = 20$)	*(histogram)* Nominal	*Ye Musket*
In control, and capable ($\sigma = 10$)	*(histogram)* Nominal	**The rifle**

Transparency Masters to Accompany *SPC Essentials and Productivity Improvement*
ASQC Quality Press   ©1997 by Harris Corporation, Semiconductor Sector

Histograms and Targets, cont.

➤ In this example, the process mean has shifted.

➤ Capable, but out of control

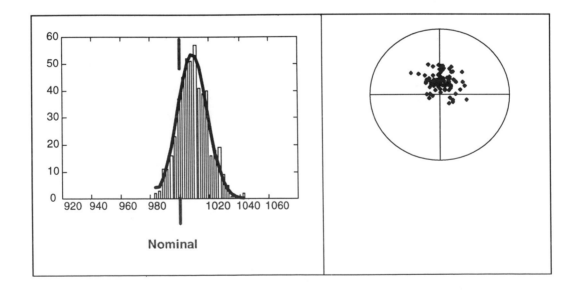

Transparency Masters to Accompany *SPC Essentials and Productivity Improvement*
ASQC Quality Press   ©1997 by Harris Corporation, Semiconductor Sector

Control Charts (Preview)

- There are SPC charts that correspond to these histograms and targets.
  - Has the variation increased?
  - Has the process mean shifted?
- We will look at these charts shortly.

Transparency Masters to Accompany *SPC Essentials and Productivity Improvement*
ASQC Quality Press   ©1997 by Harris Corporation, Semiconductor Sector

The Normality Assumption

➤ Traditional SPC relies on the assumption that the process follows the normal distribution. If the process doesn't, we must use alternative procedures.

➤ The procedures for handling non-normal distributions are beyond the scope of this course.

An engineer or statistician will handle the distributions.

The Normality Assumption, cont.

➤ A one-sided specification suggests that the process may be non-normal.

- ➤ Example: Impurities (ppm) in a process chemical
- ➤ There will be an upper limit, for example, 10 ppm.
- ➤ There can't be less than zero.

The Normality Assumption, cont.

➤ For example, a chemical, 4 ppm impurity (mean)

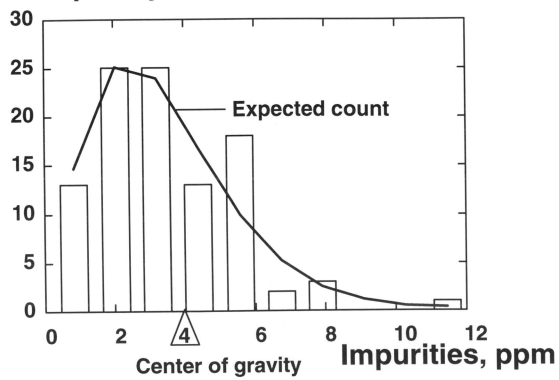

Frequency of Occurrence

Expected count

Center of gravity

Impurities, ppm

Transparency Masters to Accompany *SPC Essentials and Productivity Improvement*
ASQC Quality Press   ©1997 by Harris Corporation, Semiconductor Sector

Feedback Process Control

- ➤ Feedback process control means
  - ➤ 1. Observing data from the process
  - ➤ 2. Adjusting the process accordingly
  - ➤ 3. Observing data again to make sure that the adjustment was effective

- ➤ In 1645, Miyamoto Musashi, a samurai, wrote that a bow and arrow is better than a gun because the arrow provides feedback.

Feedback and Control Loop

Arrow: visible in flight, provides feedback to archer.

Bullet: Moves too fast to be seen.

Provides no feedback to the gunner.

Tracer bullet: Gives the gunner immediate feedback.

(Twentieth century invention)

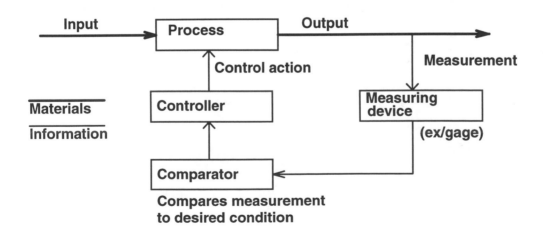

Transparency Masters to Accompany *SPC Essentials and Productivity Improvement*
ASQC Quality Press   ©1997 by Harris Corporation, Semiconductor Sector

Household Thermostat

➤ Example of feedback control

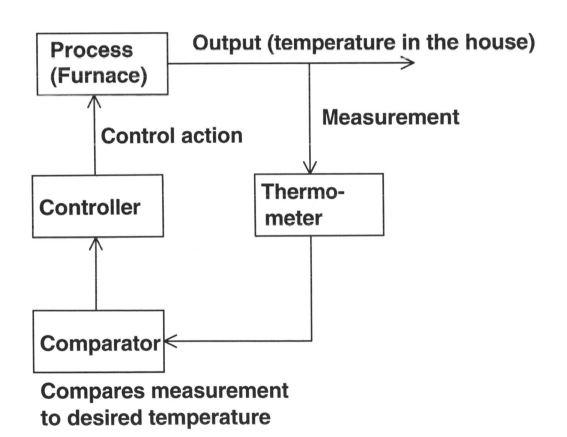

Process (Furnace) — Output (temperature in the house) →

Control action

Measurement

Controller

Thermo-meter

Comparator

Compares measurement to desired temperature

Transparency Masters to Accompany *SPC Essentials and Productivity Improvement*
ASQC Quality Press   ©1997 by Harris Corporation, Semiconductor Sector

Feedback Control: Key Aspects

➤ Measurement of the process' output

➤ Comparison of the measurement to the desired conditions

➤ Control action, if necessary, to achieve the desired conditions

   ➤ Follow-up measurement, if necessary, to verify

Transparency Masters to Accompany *SPC Essentials and Productivity Improvement*
ASQC Quality Press   ©1997 by Harris Corporation, Semiconductor Sector

Feedback Control: Key Aspects, cont.

- Archer
  - 1. Watches (measures) his arrows in flight.
  - 2. Compares them against the desired condition.
  - 3. Adjusts (controls) his aim accordingly.
- Thermostat
  - 1. Measures the temperature.
  - 2. Compares the temperature against the desired condition (set point).
  - 3. Adjusts the condition by turning the furnace or air conditioner on or off.

Transparency Masters to Accompany *SPC Essentials and Productivity Improvement*
ASQC Quality Press   ©1997 by Harris Corporation, Semiconductor Sector

Closing the Loop

➤ *Closing the loop* means acting and measuring again to confirm the effectiveness of the action.

  ➤ Measurement has little value without action.

  ➤ Action without measurement can be worse than useless!

➤ SPC control loop

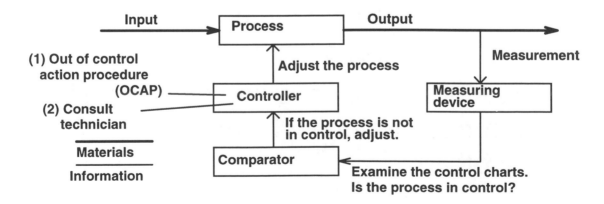

Transparency Masters to Accompany *SPC Essentials and Productivity Improvement*
ASQC Quality Press   ©1997 by Harris Corporation, Semiconductor Sector

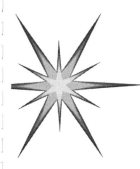

Requirements for Successful SPC

- Data integrity
- Data traceability
- Identification of critical process parameters
  - Harris Semiconductor calls these *critical nodes.*
- Real-time capability

Requirements for Successful SPC

Data integrity	Data (measurements) must be accurate. We will later examine gage capability, which is the ability of gages or instruments to make accurate and repeatable measurements.
Data traceability	This means being able to trace measurements to the processes, equipment, and material that produced them.
Identify critical process parameters	We must identify the process steps that have significant effects on product quality. Harris Semiconductor calls these critical nodes. Other companies may use other terms.
Real-time capability	Feedback must be prompt enough to allow timely process adjustments.

Transparency Masters to Accompany *SPC Essentials and Productivity Improvement*
ASQC Quality Press   ©1997 by Harris Corporation, Semiconductor Sector

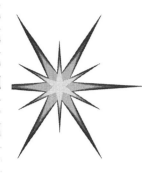

Data Traceability

➤ Which workstation produced the measurement?

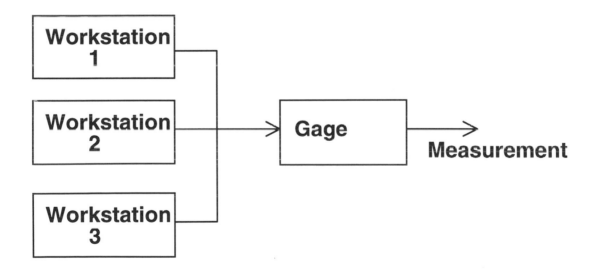

Transparency Masters to Accompany *SPC Essentials and Productivity Improvement*
ASQC Quality Press   ©1997 by Harris Corporation, Semiconductor Sector

Data Traceability, cont.

➤ Which table of process data looks more useful?

Without traceability			With traceability		
Wafer	Oxide thickness, A		Wafer	Oxide thickness, A	Oxide Workstation
8470021	975		8470021	975	2
8470350	998		8470350	998	2
8470226	1002		8470226	1002	3
8470148	982		8470148	982	2
8470223	1010		8470223	1010	1
8470104	1005		8470104	1005	3

Transparency Masters to Accompany *SPC Essentials and Productivity Improvement*
ASQC Quality Press   ©1997 by Harris Corporation, Semiconductor Sector

Real-Time Feedback

➤ Which feedback control loop is more effective?

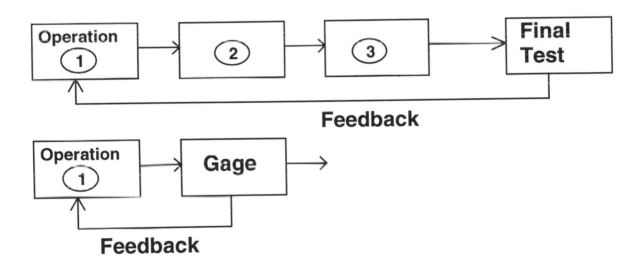

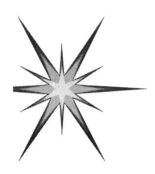

Outliers

➤ *Outliers* are unusual measurements that are out of line with the rest of the data.

➤ All outliers require investigation.

  ➤ Remeasure; the measurement may be incorrect.

  ➤ The process did something unusual or undesirable.

# Target With and Without Outlier

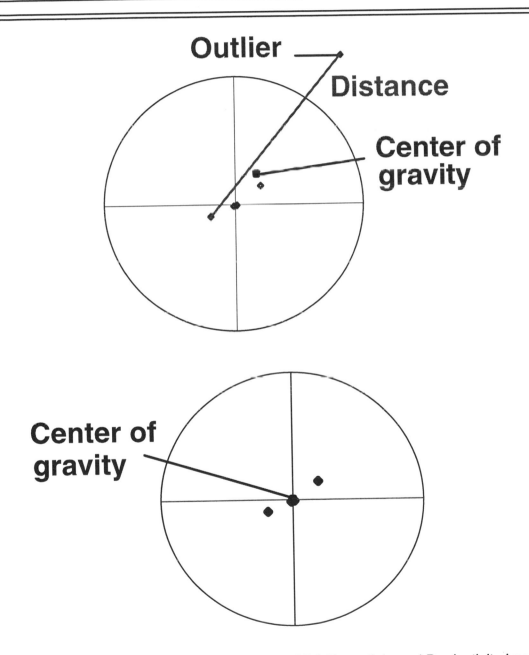

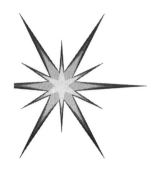

Outliers, cont.

➤ Outlier on a number line (weightless beam)

  ➤ Note the leverage on the center of gravity and the effect on the range.

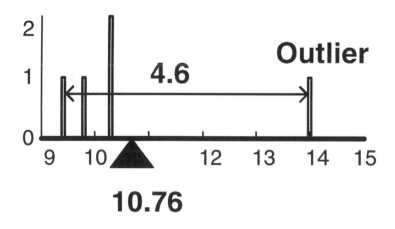

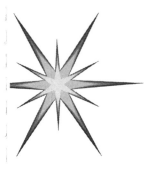

Control Charts

➤ Control charts tell us whether the manufacturing process needs adjustment.

➤ A control chart is a graphical method for assuring the consistency of a process.

Transparency Masters to Accompany *SPC Essentials and Productivity Improvement*
ASQC Quality Press   ©1997 by Harris Corporation, Semiconductor Sector

Risks in Statistical Testing

Decision, state of nature	Decide that there is a problem	Decide that there is no problem
There isn't a problem; the situation is as it should be.	False alarm risk • The boy who cried wolf • Risk of convicting an innocent defendant • Quality acceptance sampling; risk of rejecting a good lot • SPC; risk of calling the process out of control when it is in control	100% – false alarm risk • Chance of acquitting an innocent defendant • Quality acceptance sampling; chance of accepting a good lot • SPC; chance of calling the process in control when it is
There is a problem; the situation requires adjust-ment	A test's power is its ability to detect a real problem. • Chance of seeing the wolf coming • Chance of convicting a guilty defendant • *Quality acceptance sampling*; chance of rejecting a bad lot • *SPC*; chance of calling the process out of control when it is	Risk of missing the problem • Risk of not seeing the wolf • Risk of acquitting a guilty defendant • *Quality acceptance sampling*; risk of shipping a bad lot • *SPC*; risk of calling the process in control when it isn't

Transparency Masters to Accompany *SPC Essentials and Productivity Improvement*
ASQC Quality Press   ©1997 by Harris Corporation, Semiconductor Sector

Sample Size

➤ Large samples are better at telling good and bad lots, or processes, apart.

% Chance of rejecting the lot (power, sampling plan)

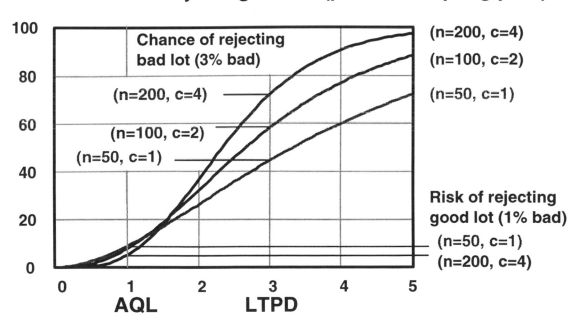

% nonconforming (bad) pieces in the lot

# Sample Size, cont.

- ➤ Plans with large samples are better at telling good lots from bad ones.

  - ➤ In SPC, large samples are better at detecting assignable causes in the process.

- ➤ It is easier to detect a big problem than a little one.

Transparency Masters to Accompany *SPC Essentials and Productivity Improvement*
ASQC Quality Press   ©1997 by Harris Corporation, Semiconductor Sector

 # Data Collection for SPC

- ➤ No statistical control method is better than the information that goes into it!

- ➤ 1. The operating instruction should include the sampling method.

  - ➤ The instruction tells how many pieces to measure.

  - ➤ The instruction tells what to measure.

- ➤ 2. Perform the measurements carefully. If the numbers don't look right, don't assume they are wrong.

  - ➤ Unusual measurements are outliers.

  - ➤ Outliers are evidence of assignable causes or problems with the gage.

Transparency Masters to Accompany *SPC Essentials and Productivity Improvement*
ASQC Quality Press   ©1997 by Harris Corporation, Semiconductor Sector

Data Collection, cont.

➤ 3. Record the results accurately.

    ➤ Be sure to assign the measurement to the process and workstation that produced it.

    ➤ Double check the numbers, especially when typing them into a computer.

Samples Should Be Random

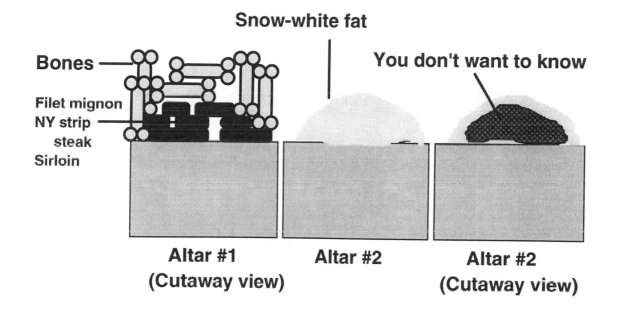

Snow-white fat

Bones

You don't want to know

Filet mignon
NY strip
steak
Sirloin

Altar #1
(Cutaway view)

Altar #2

Altar #2
(Cutaway view)

Truck trailer (cutaway view)

Loading dock

→

& Inspector

◼ **Defective**   ☐ **Good**

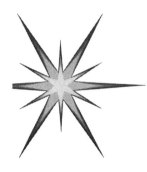

Samples Should Be Random, cont.

➤ Nonrandom sampling in a manufacturing process

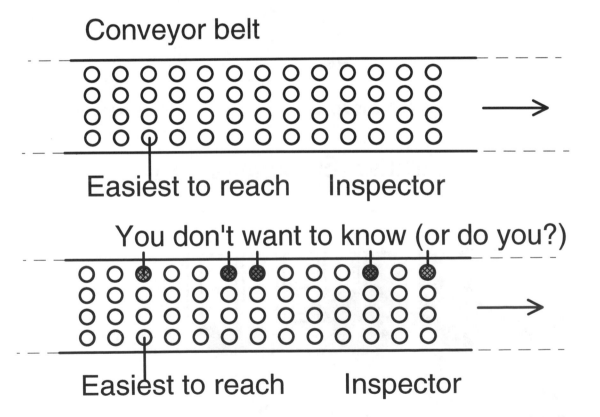

Conveyor belt

Easiest to reach   Inspector

You don't want to know (or do you?)

Easiest to reach   Inspector

Transparency Masters to Accompany *SPC Essentials and Productivity Improvement*
ASQC Quality Press   ©1997 by Harris Corporation, Semiconductor Sector

Systematic Sampling

➤ Example: Look at every fifth piece (or group).

    ➤ In the conveyor belt example, the sampling plan will eventually catch the problem.

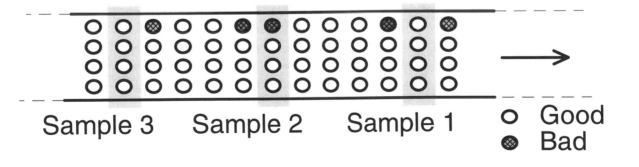

Sample 3   Sample 2   Sample 1   ◯ Good
                                               ◉ Bad

Transparency Masters to Accompany *SPC Essentials and Productivity Improvement*
ASQC Quality Press   ©1997 by Harris Corporation, Semiconductor Sector

Control Chart Symbols

Symbol	Meaning	Explanation
X	Individual measurement	In some manufacturing processes, we can get only one measurement from each sample.
$\overline{X}$ "x bar"	Sample average	For a sample of *n* measurements, add the numbers and divide by *n*.
$\overline{\overline{X}}$ "x bar bar"	Grand average of all the data	The grand average is an estimate of the actual process mean, or center of gravity. It defines the centerline of the *x*–bar control chart.
R	Sample range	Subtract the smallest number in the sample from the largest.
$\overline{R}$ "R bar"	Average range, all samples	This allows estimation of the process variation.
S	Sample standard deviation	The computer will calculate this.
UCL	Upper control limit	A point above the UCL means that the process is out of control.
LCL	Lower control limit	A point below the LCL also means that the process is out of control.
CL	Centerline	

Transparency Masters to Accompany *SPC Essentials and Productivity Improvement*
ASQC Quality Press   ©1997 by Harris Corporation, Semiconductor Sector

Control Limits

- Don't worry about where the control limits come from.
  - The chance of exceeding a control limit, if the process is in control, is 0.135%.
    - This is the chance of crying wolf.
  - When there are upper and lower limits, the total risk is 0.27%, or 2.7 per 1000.

Transparency Masters to Accompany *SPC Essentials and Productivity Improvement*
ASQC Quality Press   ©1997 by Harris Corporation, Semiconductor Sector

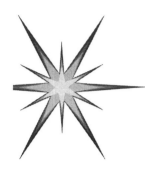 # Control Limits, cont.

➤ Here is the effect of adjusting the control limits and sample.

Action	Effect on sensitivity to real problems	Effect on false alarm rate (per sample)
Widen the control limits	Reduces sensitivity	Reduces false alarms
Tighten the control limits	Increases sensitivity	Increases false alarms
Take a larger sample	Increases sensitivity	No change

Transparency Masters to Accompany *SPC Essentials and Productivity Improvement*
ASQC Quality Press   ©1997 by Harris Corporation, Semiconductor Sector

Example: Calculations

➤ Sample average and range

Measurement	Sample 1	Sample 2	Sample 3	Sample 4	Sample 5
1	986.8	1001.9	999.6	1002.0	986.3
2	1010.9	1000.0	993.9	1005.0	993.0
3	994.9	1002.4	998.6	992.0	1009.7
4	991.0	984.5	1008.9	1010.7	998.5
5	1021.9	1005.3	1003.8	1000.9	996.6
Average	5005.5 ÷ 5 = **1001.10**	4994.1 ÷ 5 = **998.82**	5004.8 ÷ 5 = **1000.96**	5010.6 ÷ 5 = **1002.12**	4984.1 ÷ 5 = **996.82**
Range	1021.9 − 986.8 = **35.1**	1005.3 − 984.5 = **20.8**	1008.9 − 993.9 = **15.0**	1010.7 − 992.0 = **18.7**	1009.7 − 986.3 = **23.4**

Transparency Masters to Accompany *SPC Essentials and Productivity Improvement*
ASQC Quality Press   ©1997 by Harris Corporation, Semiconductor Sector

Range Chart
(for process variation)

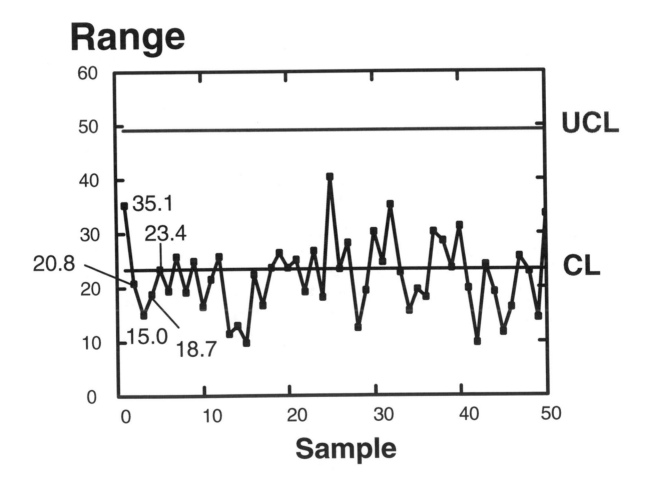

x Bar Chart (for process mean)

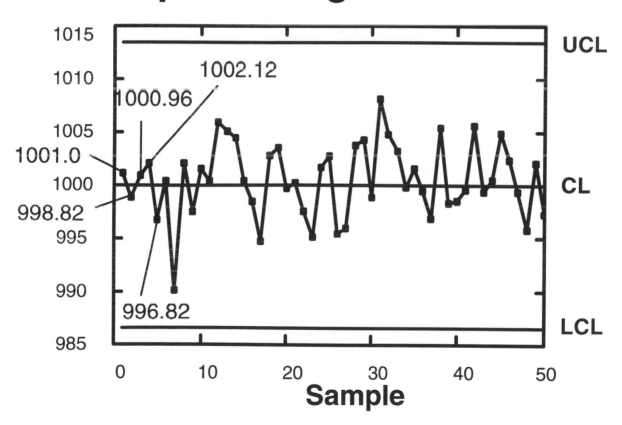

Sample Average

Transparency Masters to Accompany *SPC Essentials and Productivity Improvement*
ASQC Quality Press ©1997 by Harris Corporation, Semiconductor Sector

213

Specifications?

➤ The specification limits do not appear on the control chart.

  ➤ There is no relationship between specification and control limits.

  ➤ Products are in or out of specification.

  ➤ Processes are in or out of control.

Control Versus Specification

Situation			Product	Process
Out of control, in spec	*Out of control* USL UCL CL LCL LSL		Ship it; it meets specification.	Requires adjustment; it is out of control.
In control, out of spec	*Out of spec* UCL USL CL LSL LCL		Rework or scrap	Do nothing. The process is in control, but not capable (for example, a musket).

Transparency Masters to Accompany *SPC Essentials and Productivity Improvement*
ASQC Quality Press   ©1997 by Harris Corporation, Semiconductor Sector

Control Versus Specification, cont.

- ➤ For sample averages
    - ➤ Example: Sample of 4
    - ➤ Specification [360, 440] Angstroms
    - ➤ Control limits [385, 415]
    - ➤ Sample: 350, 400, 400, 400
    - ➤ In control, but 350 is out of spec
    - ➤ Investigate?

Interpreting Control Charts

➤ In control

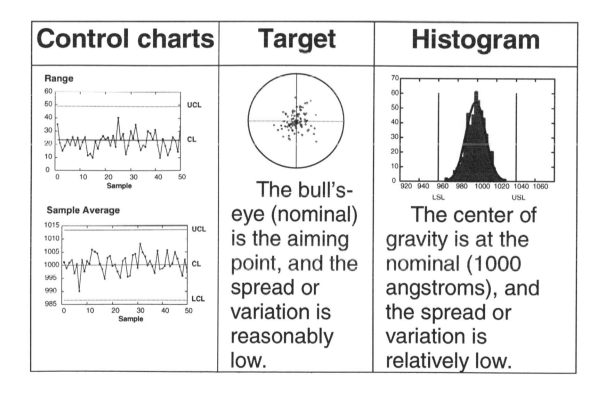

Control charts	Target	Histogram
Range [chart] Sample Average [chart]	The bull's-eye (nominal) is the aiming point, and the spread or variation is reasonably low.	The center of gravity is at the nominal (1000 angstroms), and the spread or variation is relatively low.

Transparency Masters to Accompany *SPC Essentials and Productivity Improvement*
ASQC Quality Press   ©1997 by Harris Corporation, Semiconductor Sector

Interpreting Control Charts, cont.

➤ Increase in variation

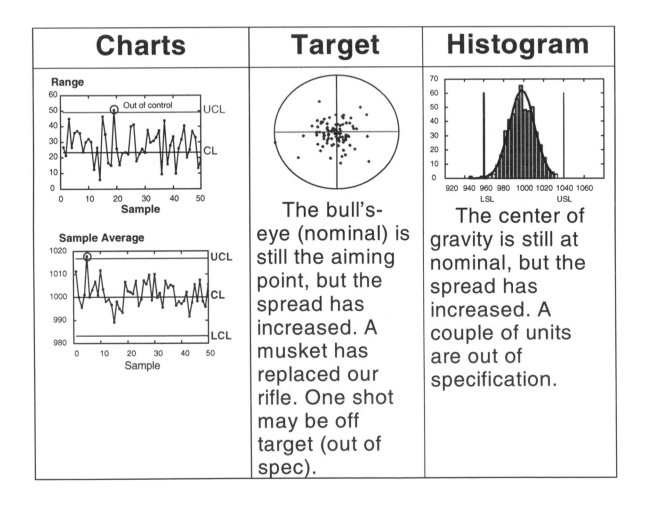

Charts	Target	Histogram
Range chart showing Out of control point near UCL, with CL and UCL lines. Sample Average chart showing point above UCL, with UCL, CL, and LCL lines.	The bull's-eye (nominal) is still the aiming point, but the spread has increased. A musket has replaced our rifle. One shot may be off target (out of spec).	The center of gravity is still at nominal, but the spread has increased. A couple of units are out of specification.

Transparency Masters to Accompany *SPC Essentials and Productivity Improvement*
ASQC Quality Press   ©1997 by Harris Corporation, Semiconductor Sector

Interpreting Control Charts, cont.

➤ Change in process mean

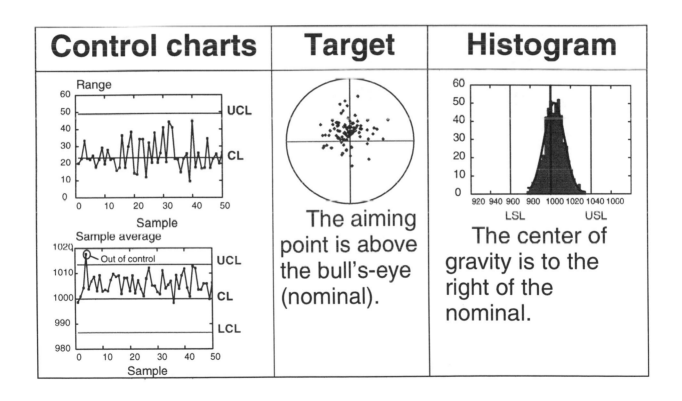

Control charts	Target	Histogram
	The aiming point is above the bull's-eye (nominal).	The center of gravity is to the right of the nominal.

Western Electric Zone Tests

- ➤ Additional tests for shifts in the process mean
- ➤ Divide the control chart into six equal zones

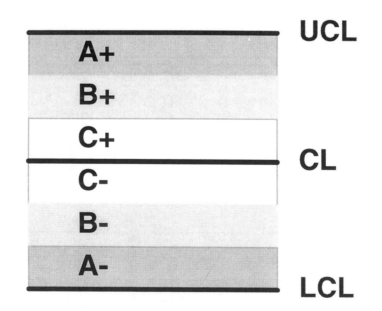

Transparency Masters to Accompany *SPC Essentials and Productivity Improvement*
ASQC Quality Press   ©1997 by Harris Corporation, Semiconductor Sector

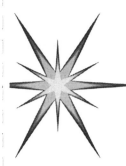

Western Electric Zone Tests, cont.

Test	Situation	Chance of false alarm
Control limits (basic)	One point exceeds either control limit.	2.7 per thousand samples
Zone A	Two out of three successive points in zone A+ *or* A–	3 per thousand
Zone B	Four out of five successive points in zone B+ *or* zone B–. (A+ counts as B+, and A– counts as B–.)	5.4 per thousand
Zone C	Eight successive points on one side of the centerline.	7.8 per thousand

Transparency Masters to Accompany *SPC Essentials and Productivity Improvement*
ASQC Quality Press   ©1997 by Harris Corporation, Semiconductor Sector

Western Electric Zone Tests, cont.

➤ Using all the zone tests increases the chance of detecting a process shift, but also increases the false alarm rate.

➤ Run of eight above or below center: 1 in 128, if the process mean is on the centerline.

  ➤ This is like throwing a coin and getting eight heads or eight tails.

H	H	H	H	H	H	H	H
1/2	1/4	1/8	1/16	1/32	1/64	1/128	1/256
T	T	T	T	T	T	T	T
1/2	1/4	1/8	1/16	1/32	1/64	1/128	1/256

2/256 = 1/128

Transparency Masters to Accompany *SPC Essentials and Productivity Improvement*
ASQC Quality Press   ©1997 by Harris Corporation, Semiconductor Sector

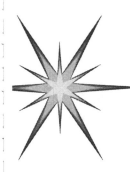

Western Electric Zone Tests, cont.

➤ 1000 Å gate oxide example

  ➤ Process mean has increased.

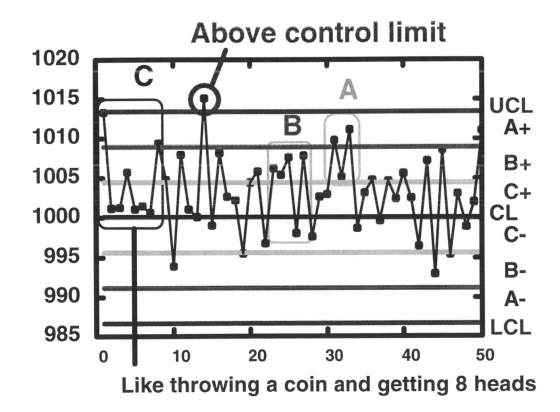

Transparency Masters to Accompany *SPC Essentials and Productivity Improvement*
ASQC Quality Press   ©1997 by Harris Corporation, Semiconductor Sector

Special Patterns

> Special control chart patterns are characteristic of

>> Certain processes

>> Incorrect control limits

> Mixture

>> This pattern may come from two different machines or material lots.

>> Always keep a separate chart for each workstation.

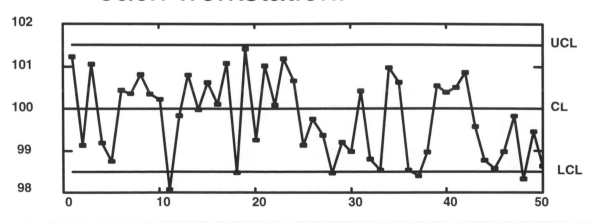

Transparency Masters to Accompany *SPC Essentials and Productivity Improvement*
ASQC Quality Press   ©1997 by Harris Corporation, Semiconductor Sector

Special Patterns, cont.

➤ Trend

  ➤ This is characteristic of a process that wears something out or uses something up.

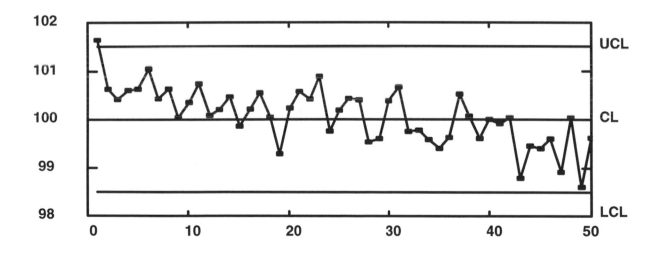

Special Patterns, cont.

- ➤ Multiple variation sources
  - ➤ This often happens in batch processes.

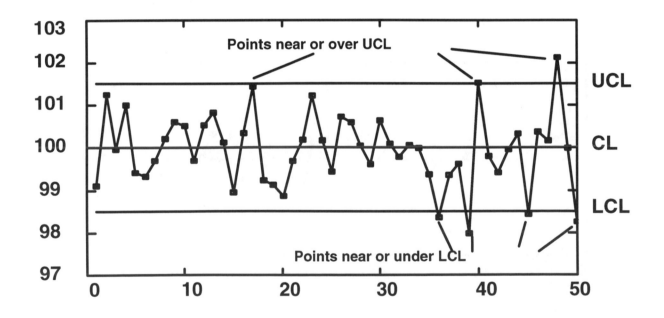

Special Patterns, cont.

➤ Stratification

   ➤ This is too good to be true.

   ➤ If an improvement reduces the process variation, recompute the limits.

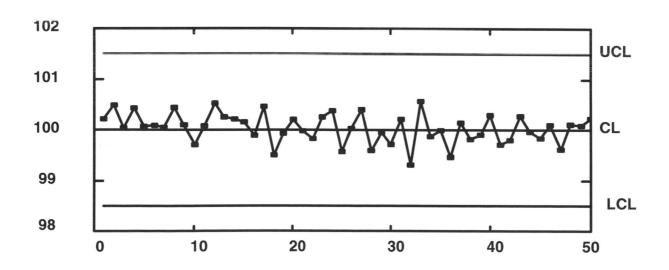

Special Patterns, cont.

➤ Overadjustment

   ➤ Note how pairs of points are on opposite sides of the centerline.

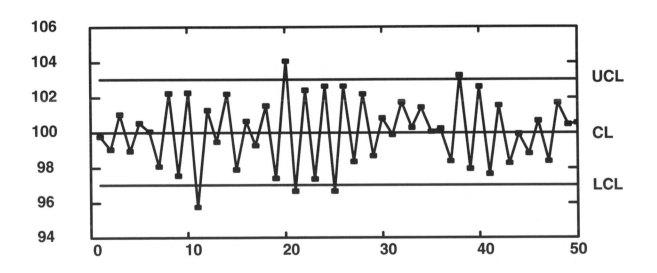

Transparency Masters to Accompany *SPC Essentials and Productivity Improvement*
ASQC Quality Press   ©1997 by Harris Corporation, Semiconductor Sector

Chapter 3, Problem 1

➤ a. Did this process need investigation and corrective action?

➤ b. If so, when should this have happened?

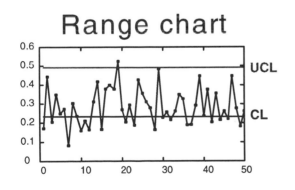

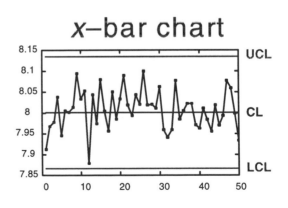

Transparency Masters to Accompany *SPC Essentials and Productivity Improvement*
ASQC Quality Press   ©1997 by Harris Corporation, Semiconductor Sector

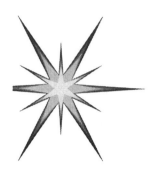

Chapter 3, Problem 2

➤ a. Did this process need investigation and corrective action?

➤ b. If so, when should this have happened?

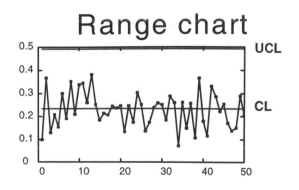

Range chart

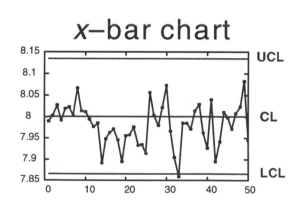

x–bar chart

Chapter 3, Problem 3

➤ a. Did the process improvement work?

➤ b. If so, when would this have been decided?

➤ c. Why do the points on the *x*-bar chart cluster around the centerline?

Range chart   x–bar chart

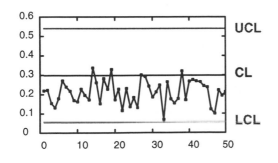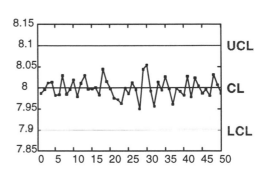

Transparency Masters to Accompany *SPC Essentials and Productivity Improvement*
ASQC Quality Press   ©1997 by Harris Corporation, Semiconductor Sector

Chapter 3, Problem 4

➤ a. Is there a problem with this process?

➤ b. If so, what is a possible explanation?

➤ c. Here is the 10th sample (measurements in mils). What action does this information suggest?

25.06 24.83 24.82 21.00 25.78

Range chart

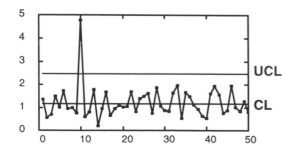

x–bar chart

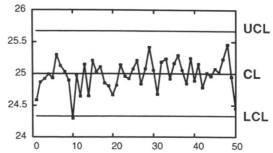

Transparency Masters to Accompany *SPC Essentials and Productivity Improvement*
ASQC Quality Press   ©1997 by Harris Corporation, Semiconductor Sector

Chapter 3, Problem 5

➤ a. What action, if any, should be taken on the process?

➤ b. What action, if any, should be taken on the 24th piece?

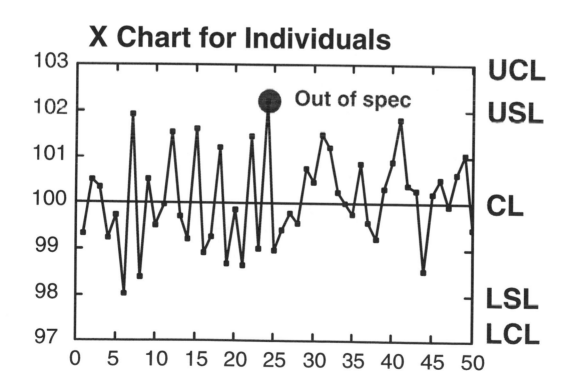

X Chart for Individuals

Attribute Control Charts

Charts for Tracking Rework, Scrap, and Defects

Transparency Masters to Accompany *SPC Essentials and Productivity Improvement*
ASQC Quality Press   ©1997 by Harris Corporation, Semiconductor Sector

Attributes

➤ Anything we must measure with whole numbers is an attribute. Attributes include

➤ Nonconformances (rejects, rework, scrap)

➤ Nonconformities (defects)

➤ Attribute control charts show when the nonconformance or defect rate has changed.

Transparency Masters to Accompany *SPC Essentials and Productivity Improvement*
ASQC Quality Press   ©1997 by Harris Corporation, Semiconductor Sector

Attributes Versus Variables

Go/No-Go Gage

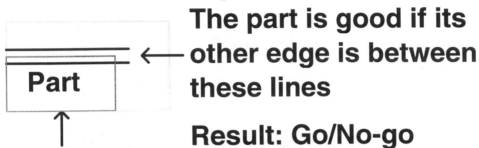

Part

Put edge of part here

The part is good if its other edge is between these lines

Result: Go/No-go
Conforming/nonconforming
Pass/reject
In spec/out of spec

Caliper or Micrometer

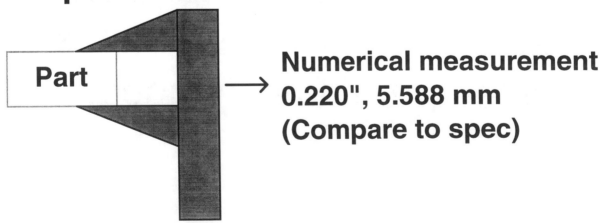

Part

Numerical measurement
0.220", 5.588 mm
(Compare to spec)

Transparency Masters to Accompany *SPC Essentials and Productivity Improvement*
ASQC Quality Press   ©1997 by Harris Corporation, Semiconductor Sector

Nonconformances and Nonconformities

- ➤ Nonconformities are defects.
- ➤ Nonconformances are rework and scrap.
  - ➤ A part can have more than one nonconformity, but it can only be one nonconformance.

Specification: 2 defects or fewer

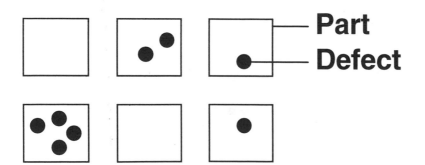

There are 8 nonconformities (defects)
There is 1 nonconformance (reject)

Transparency Masters to Accompany *SPC Essentials and Productivity Improvement*
ASQC Quality Press   ©1997 by Harris Corporation, Semiconductor Sector

Go/No-Go Data

- ➤ Gages that classify parts as good or bad are "go/no-go" gages.

- ➤ The data are binary.
  - ➤ Yes/no
  - ➤ 1/0
  - ➤ Pass/fail
  - ➤ Good/bad

- ➤ This is the least useful data we can get, but sometimes it is the only data we can get.

Transparency Masters to Accompany *SPC Essentials and Productivity Improvement*
ASQC Quality Press   ©1997 by Harris Corporation, Semiconductor Sector

Usefulness of Data

- ➤ 1. Quantitative or variables data
    - ➤ Samples, or multiple measurements, are better than individual measurements.
        - ➤ We cannot use a range or standard deviation control chart for individuals.
- ➤ 2. Nonconformity (defect) data
    - ➤ Defect counts provide more information than simple good/bad counts.
- ➤ 3. Nonconformance (good/bad, pass/fail) data
    - ➤ It's better than nothing.
- ➤ 4. No data

Transparency Masters to Accompany *SPC Essentials and Productivity Improvement*
ASQC Quality Press   ©1997 by Harris Corporation, Semiconductor Sector

Go/No-Go Data vs. *X* Chart (Individuals)

➤ Go/no-go sample with the same power as an *X* chart (individual measurement) to give out-of-control warning for the given reject rates

Go/No-Go sample size

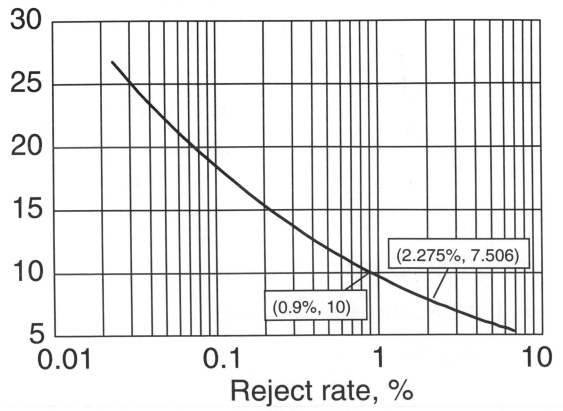

(2.275%, 7.506)

(0.9%, 10)

Reject rate, %

Variables and Attributes: Examples

Variables (quantitative data)	Attributes

• Dimensions
   • Length, width, thickness
• Temperature
• Pressure
• Relative humidity
• Electrical resistivity
   • Also current, voltage
• Viscosity
• Hardness
• Tensile strength
• Chemical concentration
• Impurity levels, for example, parts per million
	• Good/bad
	   • Rework and scrap
	   • Process yield (Good ÷ total)
	• Defect counts
	   • Defect densities
	• Particle counts
	   • For example, in a semiconductor cleanroom

Traditional Attribute Control Charts

Chart	Data Plotted	Application
np	Nonconformance (reject, scrap, rework) count, *np*	Shows whether the nonconformance rate has changed
p	Nonconformance fraction, or percent, *p*	Same
c	Defect count	Shows whether the average defect rate has changed.
u	Defect density (defects per part)	Same

np and *p* Charts (Rework/Scrap)

➤ The *np* and *p* charts detect changes in rework or scrap rates.

  ➤ They usually detect an increase, but they also can show whether a process improvement worked.

  ➤ *np* = number of nonconforming pieces

  ➤ *p* = fraction of nonconforming pieces

  ➤ *n* = sample size

Transparency Masters to Accompany *SPC Essentials and Productivity Improvement*
ASQC Quality Press   ©1997 by Harris Corporation, Semiconductor Sector

np Chart (Number Nonconforming)

➤ $n = 200$, $p = 0.02$ (On average, 2% nonconforming pieces)

Nonconforming pieces

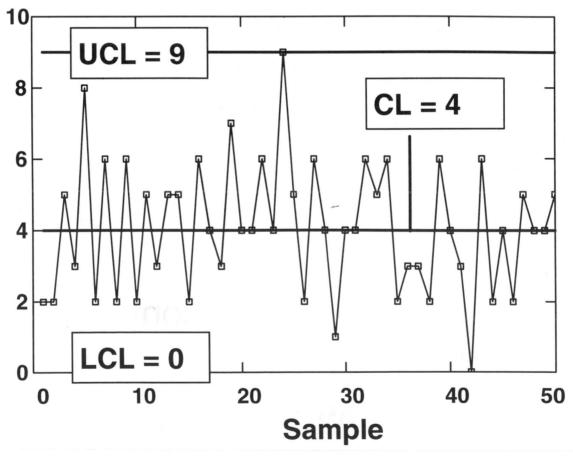

p Chart (Fraction Nonconforming)

➤ *n* = 200, *p* = 0.02 (On average, 2% nonconforming pieces)

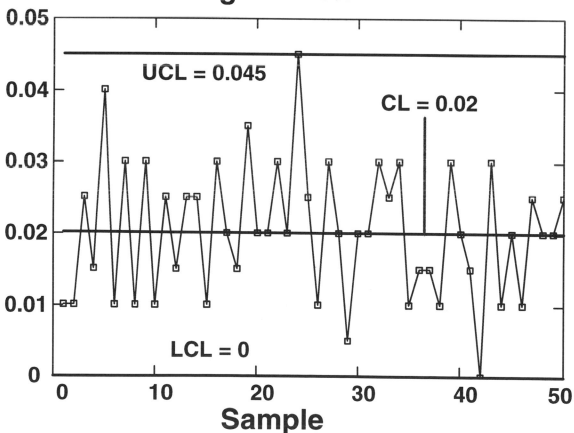

Nonconforming fraction

Transparency Masters to Accompany *SPC Essentials and Productivity Improvement*
ASQC Quality Press   ©1997 by Harris Corporation, Semiconductor Sector

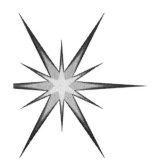

np, p, c, and *u* Charts

- ➤ A constant sample size is customary for the *np* chart, while the *p* chart accepts varying sample sizes.
  - ➤ The control limits move according to the sample size.
  - ➤ Both charts do the same job.
- ➤ The *c* and *u* charts detect changes in defect rates.
  - ➤ They also apply to other situations that involve random arrivals.
  - ➤ The *c* chart is for total defect counts, and it requires a constant sample size.
  - ➤ The *u* chart is for defect densities.

u Chart

- Here is a *u* chart for a process whose average defect density is 0.02 (2%).

- The control limits depend on the sample size.

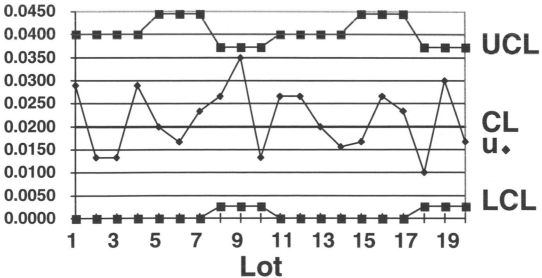

Defect density **u Chart**

Multiple Attribute Control Charts

➤ A multiple attribute control chart is a check sheet or tally sheet with control limits. Advantages include the following:

➤ They provide frequency data for Pareto charts.

➤ Their out-of-control signals point to the problem source.

➤ They are easier to use than traditional charts.

Multiple Attribute Control Charts, cont.

Samples of 50 (nonconformances)

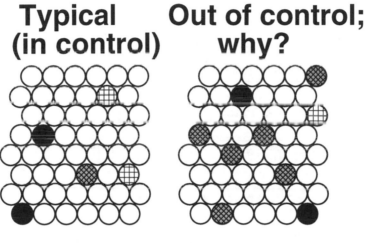

| Typical (in control) | Out of control; why? |

○ **Good**
● **Reject; problem A**
◉ **Reject; problem B**
⊕ **Reject; problem C**

➤ Traditional chart reports five and nine rejects respectively.

➤ Multiple attribute chart breaks them down by cause.

Transparency Masters to Accompany *SPC Essentials and Productivity Improvement*
ASQC Quality Press   ©1997 by Harris Corporation, Semiconductor Sector

Multiple Attribute Control Charts, cont.

➤ Consider the following process.

  ➤ Problem A has historically produced 0.5% rejects.

  ➤ Problem B produces 1.0% rejects.

  ➤ Problem C produces 0.5% rejects.

  ➤ Test a sample of 200 from each lot.

➤ Problem C increases from 0.5% to 1.0% for lots 6 through 10.

➤ A multiple attribute chart detects the out-of-control condition and points to Problem C.

➤ Spreadsheet software can handle multiple attribute charts with varying sample sizes.

Transparency Masters to Accompany *SPC Essentials and Productivity Improvement*
ASQC Quality Press   ©1997 by Harris Corporation, Semiconductor Sector

Multiple Attribute Control Charts, cont.

➤ Here is an *np* chart that tracks total rejections.

  ➤ It has the same net false alarm rate as the multiple attribute chart we will see next.

  ➤ Problem C went from 0.5% to 1.0% on the 6th sample.

  ➤ See page 252.

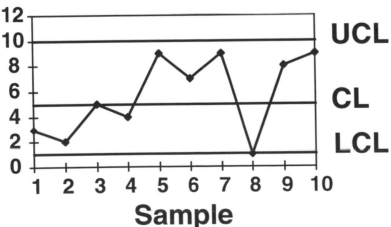

Transparency Masters to Accompany *SPC Essentials and Productivity Improvement*
ASQC Quality Press   ©1997 by Harris Corporation, Semiconductor Sector

Multiple Attribute Control Charts, cont.

➤ Multiple attribute chart

Lot	n	Nonconformance type, historical fraction			Total
		A 0.50%	**B** 1.00%	**C** 0.50%	2.00%
	UCL	4	7	4	10
	LCL	0	0	0	1
1	250	0	2	1	3
2	250	1	0	1	2
3	250	2	3	0	5
4	250	0	1	3	4
5	250	0	7	2	9
6	250	2	3	2	7
7	250	1	2	6 H	9
8	250	0	0	1	1
9	250	2	2	4	8
10	250	2	5	2	9
Total	2500	10	25	22	57
Average		0.40%	1.00%	0.88%	2.28%

Transparency Masters to Accompany *SPC Essentials and Productivity Improvement*
ASQC Quality Press   ©1997 by Harris Corporation, Semiconductor Sector

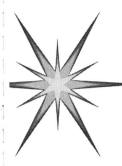

Multiple Attribute Control Charts, cont.

➤ Chart for variable sample sizes

Lot	n	Nonconformance type, historical fraction			Total	
		A 0.50%	B 1.00%	C 0.50%	2.00%	
1	250	2	2	2	6	
2	250	0	4	0	4	
3	250	2	1	0	3	
4	250	0	2	1	3	
5	200	1	2	3	6	
6	200	0	0	3	3	
7	300	0	2	2	4	
8	300	1	6	6 H	13	H
9	300	0	3	3	6	
10	300	1	4	4	9	
Total	2600	7	26	24	57	
Average		0.27%	1.00%	0.92%	2.19%	

Transparency Masters to Accompany *SPC Essentials and Productivity Improvement*
ASQC Quality Press   ©1997 by Harris Corporation, Semiconductor Sector

Technical Appendix, Chapter 4

This material describes how to set up multiple attribute charts.

Hypergeometric Distribution

- The hypergeometric distribution applies to sampling without replacement from finite populations.

    - It has some applications to acceptance sampling plans.

    - It is not relevant to SPC.

        - A process is considered an infinite population.

- When a manufacturing process is under control, nonconformances follow the binomial distribution.

Transparency Masters to Accompany *SPC Essentials and Productivity Improvement*
ASQC Quality Press   ©1997 by Harris Corporation, Semiconductor Sector

Binomial (Rework/ Scrap) Distribution

➤ The binomial distribution shows the chance of getting *x* events out of *n* trials.

   ➤ An event has a *p* chance (100*p*% chance) of happening.

➤ The second equation is the cumulative distribution.

   ➤ *c* would be the acceptance number in a sampling plan.

$$\text{Binomial } \Pr(x|n,p) = \frac{n!}{x!(n-x)!}p^x(1-p)^{n-x}$$

$$\Pr(x \le c|n,p) = \sum_{x=0}^{c} \frac{n!}{x!(n-x)!}p^x(1-p)^{n-x}$$

Transparency Masters to Accompany *SPC Essentials and Productivity Improvement*
ASQC Quality Press   ©1997 by Harris Corporation, Semiconductor Sector

Binomial Distribution, cont.

➤ Suppose we accept a 1% false alarm risk of exceeding each control limit.

$$\sum_{x=0}^{LCL-1} \frac{n!}{x!(n-x)!} p^x (1-p)^{n-x} \leq 0.01$$

$$\sum_{UCL+1}^{n} \frac{n!}{x!(n-x)!} p^x (1-p)^{n-x} \leq 0.01$$

$$\text{or} \sum_{0}^{UCL} \frac{n!}{x!(n-x)!} p^x (1-p)^{n-x} \geq 0.99$$

➤ LCL: The chance of getting $x <$ LCL is less than 1%.

➤ UCL: The chance of getting $x >$ UCL is less than 1%.

➤ The chance of getting x less than or equal to UCL is at least 99%.

Transparency Masters to Accompany *SPC Essentials and Productivity Improvement*
ASQC Quality Press   ©1997 by Harris Corporation, Semiconductor Sector

Binomial Distribution, cont.

- ➤ Hand calculations are not practical here, but spreadsheets have built-in functions for them.
  - ➤ MathCAD uses the root function.
- ➤ Spreadsheets can handle variable sample sizes and flag out-of-control points.
- ➤ The calculations use the binomial distribution, not the normal approximation.
  - ➤ Don't worry about the requirements for the normal approximation.
  - ➤ Remember, though, that small samples have little power to detect process shifts.

Transparency Masters to Accompany *SPC Essentials and Productivity Improvement*
ASQC Quality Press   ©1997 by Harris Corporation, Semiconductor Sector

Example: n = 250, p = 0.02, 3% false alarm risk at each control limit

| 0.02 | Cum Pr $(x \leq \mathbf{0} \mid 250, 0.02) = \mathbf{0.0064}$
 Cum Pr $(x \leq 1 \mid 250, 0.02) = 0.039$
 LCL = 1. There is less than a 3% chance of getting 0 rejects if n = 250 and p = 0.02. | Cum Pr $(x \leq \mathbf{10} \mid 250, 0.01) = \mathbf{0.987}$, and
 Cum Pr $(x \leq 9 \mid 250, 0.01) = 0.969$.
 UCL = 10. If the process is in control, there will be 10 or fewer nonconformances 98.7% of the time. UCL = 9 is too low, because there is a 3.1% chance of getting more than 9. |
| | = CRITBINOM (250, 0.02, 0.03) | = CRITBINOM (250, 0.02, 0.97) |

CRITBINOM is the Microsoft Excel function.

Transparency Masters to Accompany *SPC Essentials and Productivity Improvement*
ASQC Quality Press   ©1997 by Harris Corporation, Semiconductor Sector

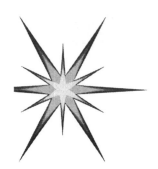

Poisson Distribution

➤ When a manufacturing process is in control, defects will follow the Poisson distribution.

  ➤ The Poisson distribution applies to random arrivals.

  ➤ The Poisson distribution also describes large samples from binomial systems, when the event probability (p) is small.

  ➤ If we expect $\mu = np$ defects, then

$$\Pr(x|\mu) = \frac{\mu^x}{x!}e^{-\mu} \quad \mu = \text{Poisson mean}$$

$$\sum_{x=0}^{c} \frac{\mu^x}{x!}e^{-\mu} = \left(\sum_{x=0}^{c} \frac{\mu^x}{x!}\right)e^{-\mu} \quad \text{(cumulative)}$$

Normal Approximation

➤ When the binomial or Poisson mean is large, the distribution behaves like a normal distribution.

  ➤ This allows us to use ±3 sigma (Shewhart) control limits for the distribution.

  ➤ The normal assumption is reasonably good when we expect four or more events (ASTM 1990, 58–59).

Source: ASTM. 1990. *Manual on presentation of data and control chart analysis*. 6th ed. Philadelphia: American Society for Testing and Materials.

Normal Approximation, cont.

	Binomial	Poisson
Mean	np	$np = \mu$
Variance (σ^2)	$np(1-p)$	μ
Standard deviation	$\sqrt{np(1-p)}$	$\sqrt{\mu}$
Shewhart limits	$np \pm 3\sqrt{np(1-p)}$	$\mu \pm 3\sqrt{\mu}$
Estimators	For p: $$\bar{p} = \frac{\sum_{i=1}^{m} x_i}{\sum_{i=1}^{m} n_i}$$ $$= \frac{\text{events}}{\text{pieces}}$$ for m samples of size n_i, with x_i nonconformances in the ith sample	The average defect count is an estimate for μ (c chart). The average defect density similar to p_bar. Instead of x_i nonconformances, use x_i defects in the ith sample.

Transparency Masters to Accompany *SPC Essentials and Productivity Improvement*
ASQC Quality Press   ©1997 by Harris Corporation, Semiconductor Sector

Other Topics

➤ Gage Capability

➤ Inspection Capability

➤ Nested Sources of Variation

Transparency Masters to Accompany *SPC Essentials and Productivity Improvement*
ASQC Quality Press   ©1997 by Harris Corporation, Semiconductor Sector

Gage Capability

- ➤ "If you can't measure it, you can't control it."

- ➤ Gages must be accurate (calibrated) and precise (capable) if they are to provide useful information.

    - ➤ An accurate gage will, on average, report the specimen's actual dimension.

    - ➤ A noncapable (imprecise) gage will return widely differing measurements from the same specimen.

    - ➤ Gage variation is similar to process variation.

Gage Capability, cont.

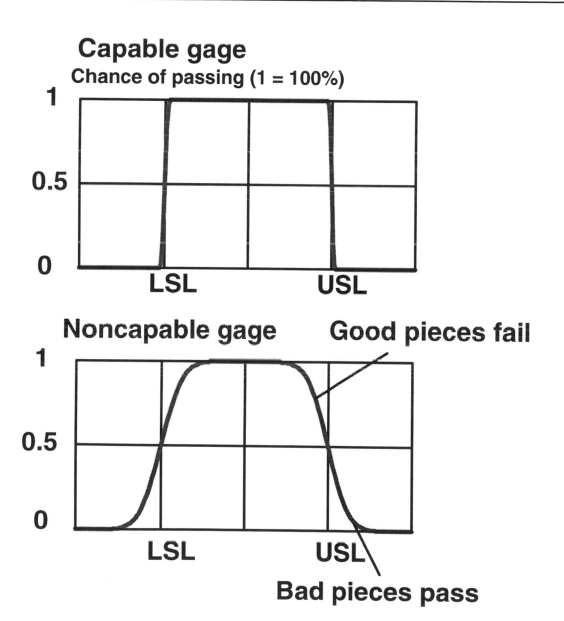

Capable gage
Chance of passing (1 = 100%)

1

0.5

0

LSL   USL

Noncapable gage   **Good pieces fail**

1

0.5

0

LSL   USL

Bad pieces pass

Transparency Masters to Accompany *SPC Essentials and Productivity Improvement*
ASQC Quality Press   ©1997 by Harris Corporation, Semiconductor Sector

Accuracy and Precision

	Accurate (calibrated)	Not accurate (out of calibration)
Precise (capable)	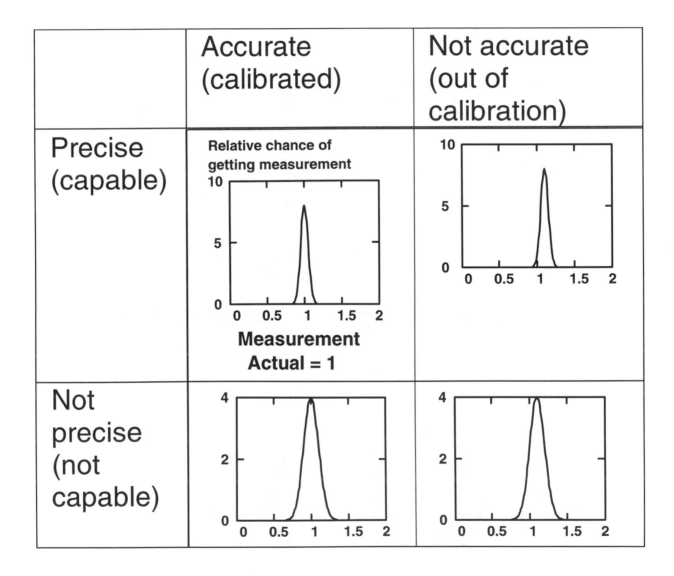	
Not precise (not capable)		

Reproducibility and Repeatability

➤ Good *reproducibility*: The measurement does not depend on the operator.

➤ Good *repeatability*: The gage returns the same number each time we measure a given specimen.

➤ A gage study, or *reproducibility and repeatability* (R&R) study, measures the gage variation.

➤ *Reproducibility* usually refers to the ability of different operators to reproduce a measurement.

Transparency Masters to Accompany *SPC Essentials and Productivity Improvement*
ASQC Quality Press   ©1997 by Harris Corporation, Semiconductor Sector

Reproducibility

➤ Why would different operators get different measurements?

Dial indicator

Mirror   **Needle reflection**

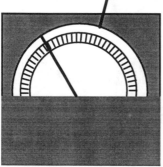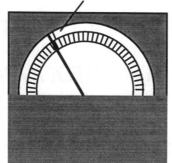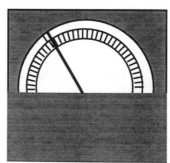

Correct: The needle covers its reflection.

Viewing position is too far to the left.

Viewing position is too far to the right.

Digital display

2 1 7 . 2 **Ohms**

Transparency Masters to Accompany *SPC Essentials and Productivity Improvement*
ASQC Quality Press   ©1997 by Harris Corporation, Semiconductor Sector

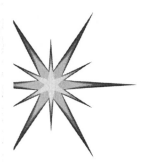

Reproducibility, cont.

> Electronic micrometer or caliper

Micrometer lines
(Place on edge of feature)

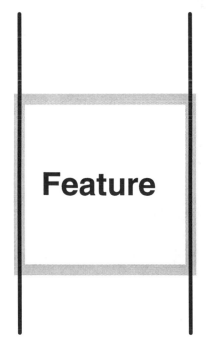

Feature

Transparency Masters to Accompany *SPC Essentials and Productivity Improvement*
ASQC Quality Press   ©1997 by Harris Corporation, Semiconductor Sector

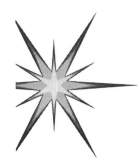

Repeatability and Total Variation

➤ *Repeatability* (or lack of repeatability) refers to variation when we measure a part several times.

    ➤ Laboratories refer to repeatability as *experimental error*.

➤ The total gage variation depends on reproducibility and repeatability as follows:

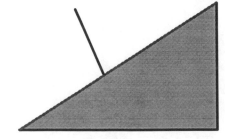

**Gage variation
(standard deviation)**

**(Lack of)
Repeatability**

(Lack of) Reproducibility

Transparency Masters to Accompany *SPC Essentials and Productivity Improvement*
ASQC Quality Press   ©1997 by Harris Corporation, Semiconductor Sector

Gage Capability

➤ Percent of tolerance consumed by (lack of) capability (PTCC)

➤ Percent of tolerance (P/T) ratio

PTCC	Status
≤ 10%	Acceptable
10–25%	Marginal
> 25%	Unacceptable

Transparency Masters to Accompany *SPC Essentials and Productivity Improvement*
ASQC Quality Press   ©1997 by Harris Corporation, Semiconductor Sector

Inspection Capability

- ➤ We cannot inspect high quality levels into a product.
    - ➤ Inspections by people are usually about 80% effective.
        - ➤ One bad item in five will get through the inspection.
- ➤ To assure high quality, we must design or build quality into the product.
    - ➤ **Design for manufacture** (DFM)
        - ➤ Design review makes sure the manufacturing process can meet the tolerances.
    - ➤ DFM builds quality into the product.

Inspection Capability, cont.

➤ Attribute inspections cannot assure extremely high quality levels.

  ➤ We would have to inspect thousands of pieces, or more.

➤ Numerical measurements (variables data) can assure high quality levels.

  ➤ A process that meets the minimum definition of *capable* makes 63.3 ppm nonconformances.

Transparency Masters to Accompany *SPC Essentials and Productivity Improvement*
ASQC Quality Press   ©1997 by Harris Corporation, Semiconductor Sector

Nested Sources of Variation

- ➤ A rational subgroup represents a homogeneous set of process conditions.

- ➤ Batch processes complicate selection of the rational subgroup.
  - ➤ Variation within batches
  - ➤ Variation between batches

Nested Sources of Variation, cont.

➤ Sequential processing; no batches

Spin coater (processes one wafer at a time)

Wafers (side view)   **Coating (thickness)**

Wafer track
⟶

This is a rational subgroup of six wafers.

Drill (processes one piece at a time)

Measure hole diameters

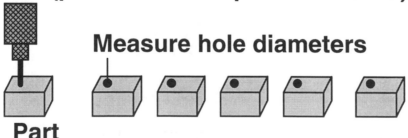

Part

This is a rational subgroup of five pieces.

Transparency Masters to Accompany *SPC Essentials and Productivity Improvement*
ASQC Quality Press   ©1997 by Harris Corporation, Semiconductor Sector

Nested Sources of Variation, cont.

➤ The batch is an independent representation of the process.

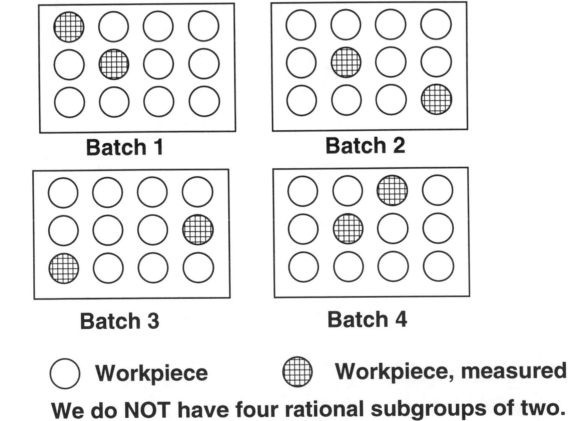

Batch 1   **Batch 2**

Batch 3   **Batch 4**

◯ **Workpiece**   ⊞ **Workpiece, measured**

We do NOT have four rational subgroups of two.
We have ONE rational subgroup of four (batches).

Nested Sources of Variation, cont.

➤ Another batch process

Tube furnace for semiconductor wafers

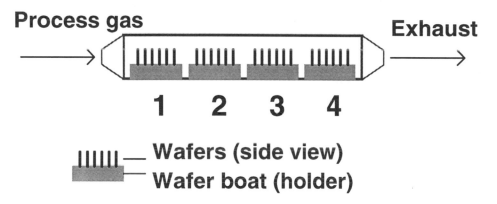

Process gas → **Exhaust** →

1   2   3   4

IIIIII — **Wafers (side view)**
▬▬▬ — **Wafer boat (holder)**

One wafer from each boat is NOT a subgroup of four.

A tube average is a subgroup of ONE.

Three tube averages make a rational subgroup of three.

Transparency Masters to Accompany *SPC Essentials and Productivity Improvement*
ASQC Quality Press   ©1997 by Harris Corporation, Semiconductor Sector

Nested Sources of Variation, cont.

➤ How can selecting the wrong subgroup cause trouble?

➤ Control limits depend on our estimate of the process variation.

Total process variation

Between-batch variation

Within-batch variation (standard deviation)

⎯⎯⎯ **Our estimate of process variation based on the wrong subgroup**

⎯⎯⎯⎯⎯ **Actual variation**

Transparency Masters to Accompany *SPC Essentials and Productivity Improvement*
ASQC Quality Press   ©1997 by Harris Corporation, Semiconductor Sector

Nested Sources of Variation, cont.

➤ If we don't account for all the variation, the control chart will look like this.

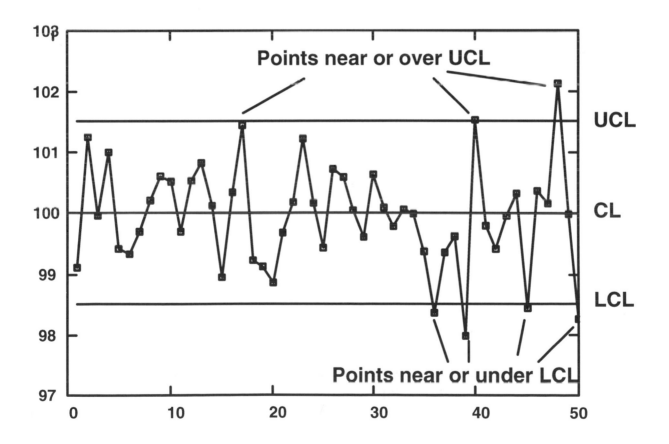

Transparency Masters to Accompany *SPC Essentials and Productivity Improvement*
ASQC Quality Press   ©1997 by Harris Corporation, Semiconductor Sector

Nested Sources of Variation, cont.

➤ Here is the same chart with the correct control limits.

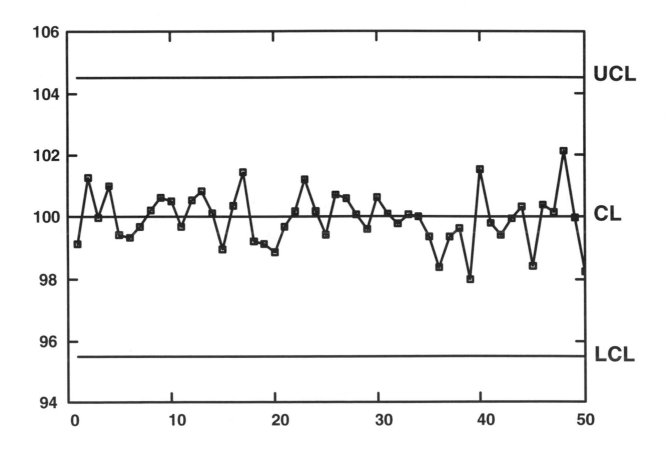

Nested Sources of Variation, cont.

➤ The chart for process variation (*R* chart or *s* chart) will look normal.

  ➤ It will, however, reflect the within-batch variation, or batch uniformity.

  ➤ It will not tell us if the overall process variation changes, that is, if the change involves between-batch variation.

Transparency Masters to Accompany *SPC Essentials and Productivity Improvement*
ASQC Quality Press   ©1997 by Harris Corporation, Semiconductor Sector

Multivariate Systems

➤ When there are systematic within-batch differences, the system can be multivariate.

➤ This is also true when we measure two or more interdependent product characteristics.

Transparency Masters to Accompany *SPC Essentials and Productivity Improvement*
ASQC Quality Press   ©1997 by Harris Corporation, Semiconductor Sector

Multivariate Systems, cont.

➤ Example: Bakery oven with hot and cold spots

	Condition of product 1 = very undercooked 2 = undercooked 3 = good 4 = overcooked 5 = burned		
	Load 1	Load 2	Load 3
Hot spot	5	4	3
Average position	4	3	2
Cold spot	3	2	1

Transparency Masters to Accompany *SPC Essentials and Productivity Improvement*
ASQC Quality Press   ©1997 by Harris Corporation, Semiconductor Sector

Multivariate Systems, cont.

➤ Tube furnace for silicon dioxide growth

   ➤ Systematic decrease in thickness from entrance to exit

Tube furnace for semiconductor wafers

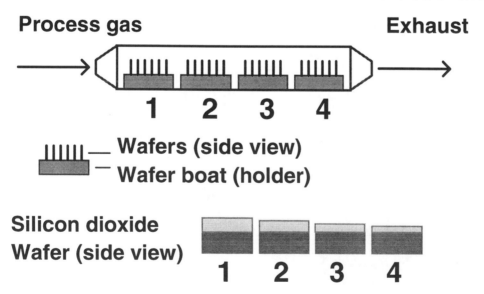

Process gas   Exhaust

1   2   3   4

⊪⊪⊪ — **Wafers (side view)**
▬▬ — **Wafer boat (holder)**

Silicon dioxide
Wafer (side view)

1   2   3   4

Transparency Masters to Accompany *SPC Essentials and Productivity Improvement*
ASQC Quality Press   ©1997 by Harris Corporation, Semiconductor Sector

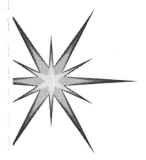

Multivariate Systems, cont.

➤ Correlation

    ➤ If the wafers near the entrance have thicker layers than usual, those near the exit will too.

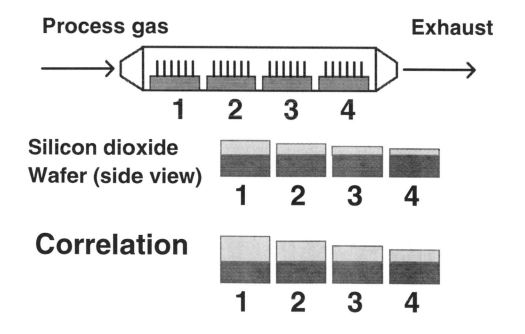

Process gas   Exhaust

1   2   3   4

Silicon dioxide
Wafer (side view)

1   2   3   4

Correlation

1   2   3   4

Transparency Masters to Accompany *SPC Essentials and Productivity Improvement*
ASQC Quality Press   ©1997 by Harris Corporation, Semiconductor Sector

Multivariate Systems, cont.

➤ Systematic differences across a process tool are usually undesirable, but are sometimes unavoidable.

➤ Bakery: Improve the oven by forcing air to circulate in it.

    ➤ Convection helps make the temperature uniform throughout the oven.

➤ When there is a correlation between two or more product characteristics, this is also a multivariate situation. These systems need special statistical techniques.

Technical Appendix, Chapter 5

Variance Components
Gage Studies

Transparency Masters to Accompany *SPC Essentials and Productivity Improvement*
ASQC Quality Press   ©1997 by Harris Corporation, Semiconductor Sector

Variance Components

Gage	Variance $\left(\sigma^2_{gage}\right)$ of one measure-ment from one specimen	$\sigma^2_M + \sigma^2_R$	Variance of n measure-ments from one specimen	$\sigma^2_M + \dfrac{\sigma^2_R}{n}$
	M = Reproducibility, R = Repeatability			
Batch pro-cess	Variance $\left(\sigma^2_{process}\right)$ of one piece from a batch	$\sigma^2_B + \sigma^2_W$	Variance of a sample of n pieces from a batch	$\sigma^2_B + \dfrac{\sigma^2_W}{n}$
	B = Between batches, W = Within batches			

Variance Components, cont.

➤ Left: Variance components for batch process; W = within, B = between

➤ Right: Variance components for gage; M = reproducibility, R = repeatability

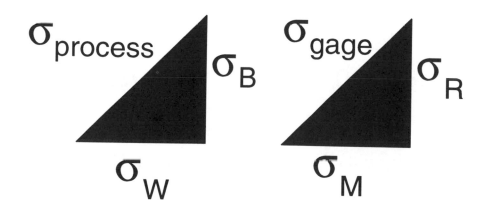

$\sigma_{process}$ σ_B σ_{gage} σ_R

σ_W σ_M

Transparency Masters to Accompany *SPC Essentials and Productivity Improvement*
ASQC Quality Press   ©1997 by Harris Corporation, Semiconductor Sector

Variance Components, cont.

➤ Batch processes: Use one-way analysis of variance (ANOVA) to isolate the variance components.

  ➤ Spreadsheets: Use built-in analysis tools to get the mean squares for error and mean squares for treatments (MSE, MST).

➤ MSE reflects within-batch variation.

➤ MST reflects within and between batch variation.

➤ The F test shows whether variation between batches is significant.

  ➤ This test is part of the built-in spreadsheet ANOVA analysis tool.

Transparency Masters to Accompany *SPC Essentials and Productivity Improvement*
ASQC Quality Press   ©1997 by Harris Corporation, Semiconductor Sector

Variance Components, cont.

➤ If there is significant between-batch variation and r replicates (measurements) per batch, then

$$\sigma^2_{within} = MSE$$

$$\sigma^2_{between} = \frac{MST - MSE}{r}$$

➤ There is an equation in the book for varying replicates per batch.

Example: Nested System ($r = 3$)

Mean for the ith run	$\mu_i = 1000 + y_i \sim N\left(0,18^2\right)$
jth piece in ith run	$x_{ij} = \mu_i + e_{ij} \sim N\left(0,10^2\right)$

$$\sigma_{within} = \sqrt{MSE} = 9.71 \text{ (vs. 10)}$$

$$\sigma_{between} = \sqrt{\frac{MST - MSE}{r}} = \sqrt{\frac{1048.5 - 94.2}{3}} = 17.84 \text{ (vs. 18)}$$

$$\sigma_{process} = \sqrt{9.71^2 + 17.84^2} = 20.31$$

$$\sigma_{Sample\ average,\ n=3} = \sqrt{17.84^2 + \frac{9.71^2}{3}} = 18.70$$

Transparency Masters to Accompany *SPC Essentials and Productivity Improvement*
ASQC Quality Press   ©1997 by Harris Corporation, Semiconductor Sector

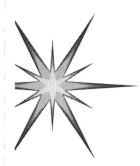

Example: Nested System, cont.

- $\sigma_{\text{sample average}} = 18.70$

  - If we take the standard deviation of the 25 averages, we get 18.69.

- Individual pieces (not samples) are in or out of spec. Therefore, σ_{process} (20.31) is the divisor for the process capability indices.

- The grand average of the 75 measurements is 1005.2, so the correct control limits for an *x*-bar chart are $1005.2 \pm 3 \times 18.7$.

  - An incorrect chart would fail to account for the variation between runs and would use $1005.2 \pm 3 \times 9.71/3^{0.5}$.

Transparency Masters to Accompany *SPC Essentials and Productivity Improvement*
ASQC Quality Press   ©1997 by Harris Corporation, Semiconductor Sector

Gage Studies

➤ Hradesky (1988), Montgomery (1991), and Barrentine (1991) describe slightly different ways of analyzing gage studies.

    ➤ We will cover Barrentine's procedure, which uses the General Motors long form. The GM procedure uses a 5.15σ interval, which contains 99% of a normal distribution, to define R&R.

    ➤ Other references use a 6σ interval, which contains 99.73% of a normal distribution.

    ➤ StatGraphics (by Manugistics) apparently uses the GM procedure.

Sources: Barrentine, Larry B. 1991. *Concepts for R&R studies*. Milwaukee: ASQC Quality Press.

Hradesky, John. 1988. *Productivity and quality improvement—A practical guide to implementing statistical process control*. New York: McGraw Hill.

Montgomery, Douglas. 1991. *Introduction to statistical quality control*. 2d ed. Milwaukee: ASQC Quality Press and New York: John Wiley & Sons.

Gage Studies, cont.

- ➤ Barrentine (1991) says to have at least two operators measure 10 or more specimens at least twice.
  - ➤ Operators ==> reproducibility component
  - ➤ Repeated measurements ==> repeatability component
- ➤ The statistical model for gage components is as follows:

	Equation
Bias for the j th operator	$\tau_j \sim N\left(0, \sigma_M^2\right)$
k th measurement by j th operator, i th specimen	$x_{ijk} = d_i + \tau_j +$ $e_{ijk} \sim N\left(0, \sigma_R^2\right)$

Source: Barrentine, Larry B. 1991. *Concepts for R&R studies*. Milwaukee: ASQC Quality Press.

Gage Studies, cont.

➤ GM long form (two of four operators)

Spec-imen	Operator 1				Operator 2			
	Rep. 1	Rep. 2	Rep. 3	R	Rep. 1	Rep. 2	Rep. 3	R
1	98.45	105.61	100.07	7.16	98.97	99.42	97.43	1.99
2	106.17	112.25	108.60	6.08	98.96	105.66	103.61	6.70
3	102.63	105.01	99.62	5.39	97.67	90.55	99.06	8.51
4	107.04	97.10	103.59	9.94	96.38	101.57	89.32	12.25
5	105.58	100.29	109.25	8.96	97.98	101.29	94.71	6.58
6	108.15	112.76	104.54	8.22	101.71	108.47	101.43	7.04
7	99.27	102.75	101.91	3.48	94.52	93.02	89.40	5.12
8	101.10	107.61	105.49	6.51	98.69	93.97	88.14	10.55
9	107.75	107.26	111.97	4.71	108.78	105.81	106.65	2.97
10	101.31	105.26	103.89	3.95	93.78	98.42	99.04	5.26
	Avg.	104.74	Avg. R	6.44	Avg.	98.48	Avg. R	6.70

Transparency Masters to Accompany *SPC Essentials and Productivity Improvement*
ASQC Quality Press   ©1997 by Harris Corporation, Semiconductor Sector

Gage Studies, cont.

➤ GM long form, cont.

Operator	Average range	Operator Average
1	6.44	104.74
2	6.70	98.48
3	6.53	97.89
4	6.50	100.25
	Grand average R = 6.54	Range = 104.74 − 97.89 = 6.85

$$\sigma_R^2 \qquad\qquad \sigma_M^2$$

Transparency Masters to Accompany *SPC Essentials and Productivity Improvement*
ASQC Quality Press   ©1997 by Harris Corporation, Semiconductor Sector

Gage Studies, cont.

➤ Formulas

	Formula
Repeat-ability	$$\sigma_R = \dfrac{\overline{\overline{R}}}{d_{2,\ k \text{ replicates}}}$$
Repro-ducibility	$$\sigma_M = \sqrt{\left(\dfrac{R_{avgs}}{d_2} c_4\right)^2 - \dfrac{\sigma_R^2}{nr}}$$ for n specimens. d_2 and c_4 are for r operators.
Gage	$$\sigma_{gage} = \sqrt{\sigma_R^2 + \sigma_M^2}$$
PTCC or P/T ratio	$$PTCC = \dfrac{5.15\sigma_{gage}}{USL - LSL}$$

Transparency Masters to Accompany *SPC Essentials and Productivity Improvement*
ASQC Quality Press   ©1997 by Harris Corporation, Semiconductor Sector

Gage Studies, cont.

Example	Simulator used:
$\sigma_R = \dfrac{6.54}{1.693} = 3.86$	4
$\sigma_M = \sqrt{\left(\dfrac{6.85}{2.059} 0.9213\right)^2 - \dfrac{3.86^2}{10 \times 4}}$ $= 3.00$	3
$\sigma_{gage} = \sqrt{3.86^2 + 3.00^2} = 4.89$	(5)
$PTCC = \dfrac{5.15 \times 4.89}{125 - 75} = 0.504$ or 50.4%	

Transparency Masters to Accompany *SPC Essentials and Productivity Improvement*
ASQC Quality Press   ©1997 by Harris Corporation, Semiconductor Sector

Gage Studies, cont.

➤ A range chart shows whether any of the ranges are unusual. Set it up like a standard range chart.

  ➤ The UCL is D_4 times the grand average range. D_4 is another control chart factor that depends on the sample size.

  ➤ Here, UCL = 16.83

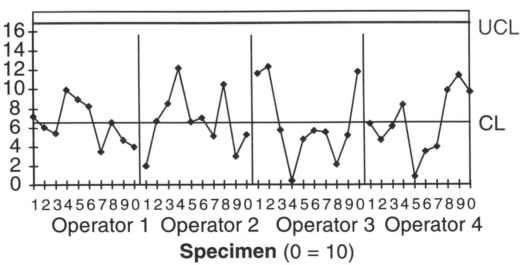

R   **R chart for gage study**

Operator 1   Operator 2   Operator 3   Operator 4

Specimen (0 = 10)

Transparency Masters to Accompany *SPC Essentials and Productivity Improvement*
ASQC Quality Press   ©1997 by Harris Corporation, Semiconductor Sector

Gages and Process Capability

- The process capability index measures the process' ability to meet the specification.

- The usual estimate for process variation includes the gage variation.

  - This reduces the estimated process capability indices.

Transparency Masters to Accompany *SPC Essentials and Productivity Improvement*
ASQC Quality Press   ©1997 by Harris Corporation, Semiconductor Sector

Gages and Process Capability, cont.

➤ The goal is to separate the gage variation component from the capability calculation.

$$\sigma_{measurement} = \sqrt{\sigma^2_{process} + \sigma^2_{gage}}$$

Using C_p as an example:

We want $C_p = \dfrac{USL - LSL}{6\sigma_{process}}$

and not $C_p = \dfrac{USL - LSL}{6\sigma_{measurement}}$

Transparency Masters to Accompany *SPC Essentials and Productivity Improvement*
ASQC Quality Press   ©1997 by Harris Corporation, Semiconductor Sector

Gages and Process Capability, cont.

➤ Barrentine (1991, 41–43) defines the actual, or inherent, process capability as follows: C_{po} is the observed process capability estimate; that is, $(USL - LSL)/(6\sigma_{measurement})$.

$$C_{pA} = \frac{1}{6\sqrt{\left(\dfrac{1}{6C_{po}}\right)^2 - \left(\dfrac{PTCC}{5.15}\right)^2}}$$

Source: Barrentine, Larry B. 1991. *Concepts for R&R studies*. Milwaukee: ASQC Quality Press.

Transparency Masters to Accompany *SPC Essentials and Productivity Improvement*
ASQC Quality Press   ©1997 by Harris Corporation, Semiconductor Sector

Gages and Process Capability, cont.

➤ Example: Observed C_p = 1.4, PTCC = 30%, actual C_p = 1.605

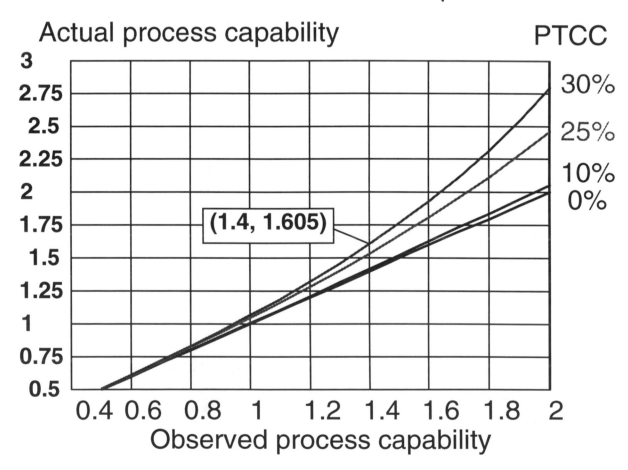

Actual process capability — PTCC

Observed process capability

Graph points: (1.4, 1.605); PTCC curves labeled 30%, 25%, 10%, 0%

Transparency Masters to Accompany *SPC Essentials and Productivity Improvement*
ASQC Quality Press   ©1997 by Harris Corporation, Semiconductor Sector

Gages and Process Capability, cont.

➤ A mediocre gage can significantly reduce the process capability estimate for a good process.

  ➤ When the process capability is poor, however, even a 30% PTCC has little effect on the estimate.

➤ A noncapable process produces many nonconformances.

  ➤ A bad gage will allow some to pass, so we must replace the gage or improve the process.

Transparency Masters to Accompany *SPC Essentials and Productivity Improvement*
ASQC Quality Press   ©1997 by Harris Corporation, Semiconductor Sector

Gages and Process Capability, cont.

➤ When the process capability is good, there will be few nonconformances (< 2 ppb when $C_{pk} = 2$).

   ➤ If such a process is in control, a new gage is not urgent.

➤ The chance of making a piece with dimension x, and getting measurement y, is as follows:

   ➤ The status of the piece depends on x. The decision to accept or reject it depends on y.

$$f(x, y) = \frac{1}{2\pi\sigma_p\sigma_g} \exp\left[-\frac{1}{2\pi}\left(\left(\frac{x - \mu_p}{\sigma_p}\right)^2 + \left(\frac{y - x}{\sigma_g}\right)^2\right)\right]$$

Transparency Masters to Accompany *SPC Essentials and Productivity Improvement*
ASQC Quality Press   ©1997 by Harris Corporation, Semiconductor Sector

Gage and Outgoing Quality

➤ The process standard deviation is σ_p, that for the gage is σ_g, and the process mean is μ_p.

➤ Double integration of this expression yields the shipments of bad products, rejections of good units, and so on.

Transparency Masters to Accompany *SPC Essentials and Productivity Improvement*
ASQC Quality Press   ©1997 by Harris Corporation, Semiconductor Sector

Gage and Outgoing Quality, cont.

Measurement from gage

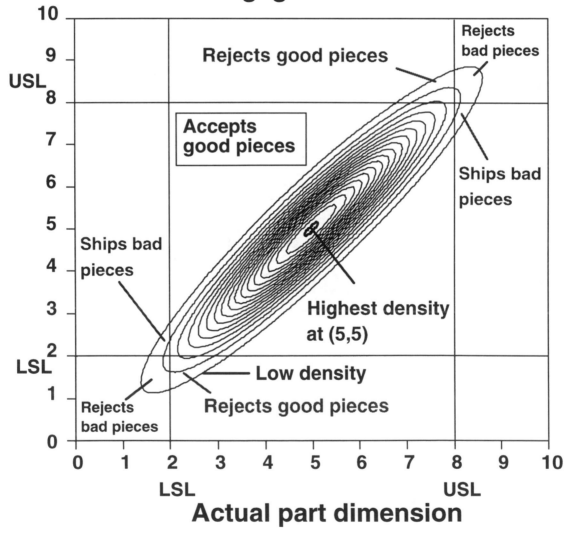

Process Characterization and Advanced Techniques

Transparency Masters to Accompany *SPC Essentials and Productivity Improvement*
ASQC Quality Press   ©1997 by Harris Corporation, Semiconductor Sector

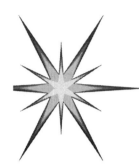

Process Characterization

➤ Introducing a control chart requires process characterization.

  ➤ We must estimate the process' mean and standard deviation.

  ➤ We must test the assumption that the process follows a normal distribution.

➤ The normal distribution is a good model for most manufacturing processes.

➤ We can use it to calculate the process yield and the chance of getting a point outside the control limits.

The Normal Distribution

- The normal distribution is a continuous distribution.

  - Its arguments are real numbers.

  - Its probability density function (pdf) shows the relative chance of getting a number from a population.

- Compare the normal distribution to discrete distributions like the binomial and Poisson.

  - Their arguments are discrete (integers).

  - Their pdfs return the chance of getting exactly x events, where x is an integer.

Transparency Masters to Accompany *SPC Essentials and Productivity Improvement*
ASQC Quality Press   ©1997 by Harris Corporation, Semiconductor Sector

The Normal Distribution, cont.

- In a continuous pdf, there is a differential or very tiny chance of getting a particular number.
  - Example: $\mu = 5$, $\sigma = 0.5$
  - The most likely result (mode) is 5.
  - There is no chance of getting exactly 5.
  - There is a 68.27% chance of getting between 4.5 and 5.5 [4.5, 5.5].
- Why should the chance of getting x in the range [a, b] interest us?
  - a and b might be the specification limits [LSL, USL].
  - a and b might be the control limits [LCL, UCL].

Transparency Masters to Accompany *SPC Essentials and Productivity Improvement*
ASQC Quality Press   ©1997 by Harris Corporation, Semiconductor Sector

The Normal Distribution, cont.

➤ Two parameters define a normal distribution.

➤ The mean, or location parameter, defines the center of gravity.

  ➤ Its symbol is the Greek letter mu (μ).

➤ Its standard deviation, or shape parameter, defines the variation or spread.

  ➤ Its symbol is the Greek letter sigma (σ).

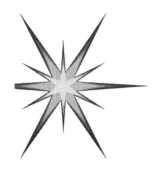

The Normal Distribution, cont.

- ➤ Greek letters symbolize actual or true values.

- ➤ English letters refer to values from samples.

- ➤ A Greek letter with a caret (^) or hat over it is an estimate.

Parameter	Population	Estimate	Sample
Mean	μ	$\hat{\mu}$	$\bar{x}$
Standard deviation	σ	$\hat{\sigma}$	s

Transparency Masters to Accompany *SPC Essentials and Productivity Improvement*
ASQC Quality Press   ©1997 by Harris Corporation, Semiconductor Sector

The Normal Distribution, cont.

➤ The normal probability density function defines the relative chance of getting value x from a normal distribution with mean μ and standard deviation σ.

  ➤ The shorthand for "x comes from a normal distribution with mean μ and variance σ^2" is $x \sim N(\mu, \sigma^2)$.

Transparency Masters to Accompany *SPC Essentials and Productivity Improvement*
ASQC Quality Press   ©1997 by Harris Corporation, Semiconductor Sector

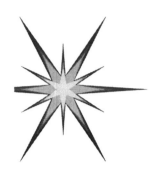

The Normal Distribution, cont.

➤ Normal pdf:

$$f(x) = \frac{1}{\sigma\sqrt{2\pi}} \exp\left(-\frac{1}{2}\left(\frac{x-\mu}{\sigma} \right)^2 \right)$$

Mode = most likely; Median = 50th percentile

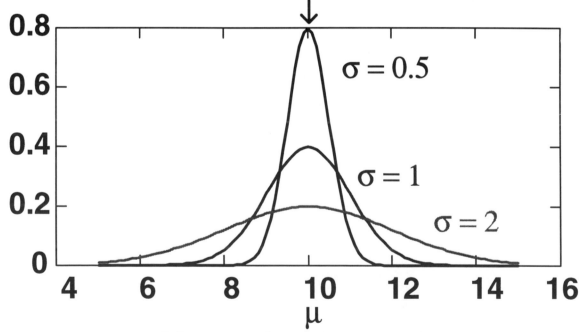

Mean: Center of gravity

Transparency Masters to Accompany *SPC Essentials and Productivity Improvement*
ASQC Quality Press   ©1997 by Harris Corporation, Semiconductor Sector

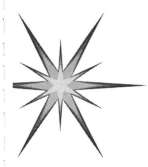

The Normal Distribution, cont.

- ➤ The mean is the most likely value, or *mode*.

- ➤ The curves are symmetric around the mean. This makes the mean the *median* (50th percentile).

- ➤ The standard deviation defines the curve's shape.

  - ➤ A high standard deviation results in a broad function. The curve for a low standard deviation is narrow.

Transparency Masters to Accompany *SPC Essentials and Productivity Improvement*
ASQC Quality Press   ©1997 by Harris Corporation, Semiconductor Sector

The Cumulative Normal Distribution

➤ The *cumulative normal distribution* lets us find the chance of getting a value within a range of numbers, [*a, b*].

$$\Pr(a \le x \le b) = \int_a^b \frac{1}{\sigma\sqrt{2\pi}} \exp\left(-\frac{1}{2}\left(\frac{y-\mu}{\sigma}\right)^2\right) dy$$

$\Pr(a \le x \le b) =$

"probability that x is in the range [a, b]."

$\int_a^b f(y)dy =$ area under

function f in range [a, b].

Cumulative Normal Distribution, cont.

➤ We can simplify this by using the *standard normal deviate, z.*

➤ *x* is *z* standard deviations from the population mean.

   ➤ There are standard tables of, and software functions for, the cumulative standard normal distribution $\Phi(z)$.

Transparency Masters to Accompany *SPC Essentials and Productivity Improvement*
ASQC Quality Press   ©1997 by Harris Corporation, Semiconductor Sector

Cumulative Normal Distribution, cont.

$$z = \frac{x - \mu}{\sigma} \quad \text{or} \quad x = \mu + z\sigma$$

(x is z standard deviations from the mean)

$$\Phi(z) = \int_{-\infty}^{z} \frac{1}{\sqrt{2\pi}} \exp\left(-\frac{1}{2}y^2\right) dy$$

Also, $\Phi(-z) = 1 - \Phi(z)$

➤ There is no closed form for the integral $\Phi(z)$.

   ➤ There are, however, tables and software functions for it.

Cumulative Normal Distribution, cont.

- ➤ What happens if *z* is negative?
  - ➤ Tables show *F(z)* only for positive *z*.
- ➤ The area under any pdf is 1 (100% of the population).
- ➤ The normal pdf is symmetric.
- ➤ Therefore, $\Phi(-z) = 1 - \Phi(z)$.

Transparency Masters to Accompany *SPC Essentials and Productivity Improvement*
ASQC Quality Press   ©1997 by Harris Corporation, Semiconductor Sector

# Cumulative Normal Distribution, cont.

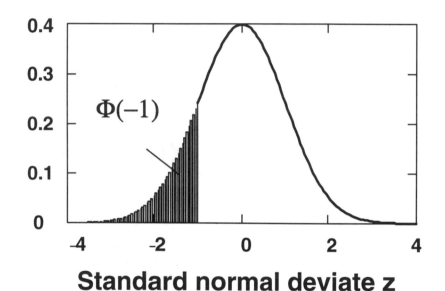

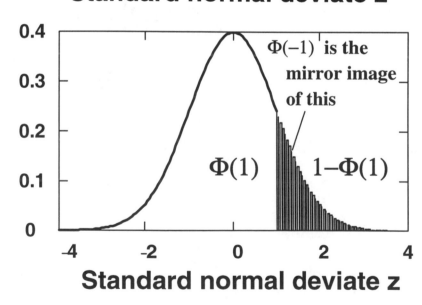

Transparency Masters to Accompany *SPC Essentials and Productivity Improvement*
ASQC Quality Press   ©1997 by Harris Corporation, Semiconductor Sector

Cumulative Normal Distribution, cont.

➤ A drill press makes holes whose nominal diameter is 1/8 inch (125 mils).

  ➤ The specification is [123, 127] mils; the process mean is 124.5 mils; and the standard deviation is 1 mil.

  ➤ What fraction will be in specification?

Transparency Masters to Accompany *SPC Essentials and Productivity Improvement*
ASQC Quality Press   ©1997 by Harris Corporation, Semiconductor Sector

 # Cumulative Normal Distribution, cont.

Portion	Standard normal deviate	$\Phi(z)$
Below the USL	$z = \dfrac{(127-124.5)\text{mils}}{1 \text{ mil}}$ $= 2.5$	$\Phi(2.5) = 0.993790.$ $1-0.99379 = 0.00621$ is above the USL.
Below the LSL	$z = \dfrac{(123-124.5)\text{mils}}{1 \text{ mil}}$ $= -1.5$	$\Phi(-1.5) = 1-\Phi(1.5)$ $=1 - 0.93319$ $= 0.06681$
Summary	0.99379 (below USL) − 0.06681 (below LSL) = 0.92698 92.698% of the product is in specification. 6.681% is below the LSL. 0.621% is above the USL.	

Transparency Masters to Accompany *SPC Essentials and Productivity Improvement*
ASQC Quality Press   ©1997 by Harris Corporation, Semiconductor Sector

Cumulative Normal Distribution, cont.

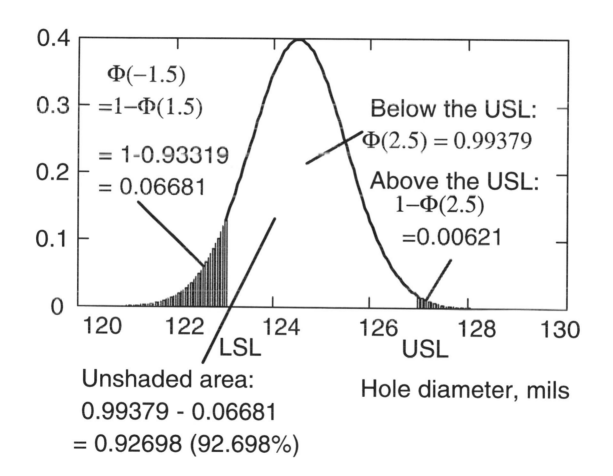

$\Phi(-1.5)$
$=1-\Phi(1.5)$

$= 1-0.93319$
$= 0.06681$

Below the USL:
$\Phi(2.5) = 0.99379$

Above the USL:
$1-\Phi(2.5)$
$=0.00621$

LSL

USL

Unshaded area:
0.99379 - 0.06681
= 0.92698 (92.698%)

Hole diameter, mils

Transparency Masters to Accompany *SPC Essentials and Productivity Improvement*
ASQC Quality Press   ©1997 by Harris Corporation, Semiconductor Sector

Cumulative Normal Distribution, cont.

➤ This process is 0.5 mil off nominal.

➤ What happens if we increase the mean to 125 mils?

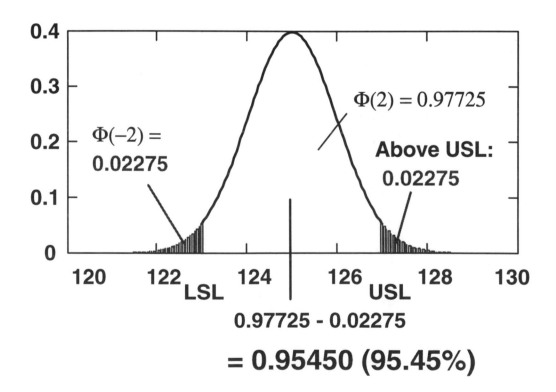

Cumulative Normal Distribution, cont.

➤ What is the chance of getting a point outside the Shewhart three-sigma control limits if the process is in control?

- ➤ Each control limit is three standard deviations from the mean.
- ➤ The chance of being below the UCL is $\Phi(3) = 0.998650$, so the chance of being over the UCL is 0.00135 (0.135%).
- ➤ The chance of being below the LCL is $\Phi(-3) = 1 - \Phi(3) = 0.00135$ (0.135%).
- ➤ The chance of being outside [LCL, UCL] is 0.270%, or 2.7 chances per 1000 samples.

Transparency Masters to Accompany *SPC Essentials and Productivity Improvement*
ASQC Quality Press   ©1997 by Harris Corporation, Semiconductor Sector

Tests for Normality

➤ Standard methods for SPC rely on the assumption that the process follows a normal distribution.

  ➤ We must check this assumption before setting up control charts.

➤ The *chi square test* quantitatively assesses a histogram's fit to a statistical distribution.

  ➤ It requires at least 30, and preferably more than 100, data.

  ➤ The results depend on the user's selection of the histogram cells.

➤ The *normal probability plot* assesses normality qualitatively and quantitatively.

# Chi Square Test

- ➤ The histogram should have a bell shape if the population follows a normal distribution.

  - ➤ The criterion "looks like a bell" is subjective.

  - ➤ The chi square test returns a statistic, χ^2 (chi square), that measures the discrepancy between the histogram and the normal distribution.

  - ➤ The chi square test is also useful for checking data for conformance to other distributions. For example, nonconformance (rework/scrap) data should follow the binomial distribution if the process is in control.

Transparency Masters to Accompany *SPC Essentials and Productivity Improvement*
ASQC Quality Press   ©1997 by Harris Corporation, Semiconductor Sector

Chi Square Test, cont.

- ➤ Prepare a histogram of the data.
    - ➤ For n data, start with $n^{0.5}$ cells (Messina 1987), or use Shapiro's (1986) equation.
    - ➤ n should be at least 30, and preferably 100 or more (Juran and Gryna 1988).
    - ➤ Count the observations in each cell. Call this f_i (frequency in cell i) or O_i (observations in cell i).
    - ➤ Compute the expected number of observations in each cell.
    - ➤ The null hypothesis (assumption) is that the data follow the normal distribution $N(\mu, \sigma^2)$.
    - ➤ μ and σ^2 may be standards or givens. Alternately, we may use the data to estimate μ and σ^2.

Sources: Juran, Joseph, and Frank Gryna. 1988. *Juran's quality control handbook*. 4th ed. New York: McGraw-Hill.

Messina, William. 1987. *Statistical quality control for manufacturing managers*. New York: John Wiley & Sons.

Shapiro, Samuel S. 1986. *How to test normality and other distributional assumptions*. Milwaukee: ASQC Quality Press.

Transparency Masters to Accompany *SPC Essentials and Productivity Improvement*
ASQC Quality Press   ©1997 by Harris Corporation, Semiconductor Sector

Chi Square Test, cont.

➤ L_i is the lower limit for the ith cell, and U_i is the upper limit. For *n* data,

$$E_i = n\int_{L_i}^{U_i} f(x)dx = n\left[\Phi\left(\frac{U_i - \mu}{\sigma} \right) - \Phi\left(\frac{L_i - \mu}{\sigma} \right) \right]$$

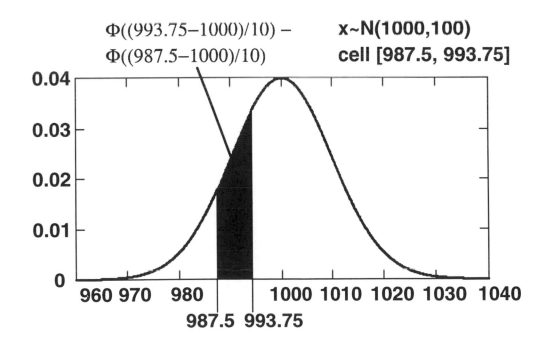

$\Phi((993.75-1000)/10) -$
$\Phi((987.5-1000)/10)$
x~N(1000,100)
cell [987.5, 993.75]

Transparency Masters to Accompany *SPC Essentials and Productivity Improvement*
ASQC Quality Press   ©1997 by Harris Corporation, Semiconductor Sector

Chi Square Test, cont.

➤ The theory behind the chi square test relies on the assumption that E_i is 5 or greater in each cell.

  ➤ If this is not true, combine cells to meet this condition.

  ➤ This usually happens in the tail areas.

➤ Compute the chi square test statistic.

  ➤ χ^2 is the evidence against normality.

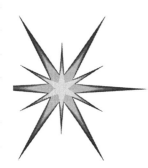

Chi Square Test, cont.

$$\chi^2 = \sum_{i=1}^{k} \frac{(O_i - E_i)^2}{E_i}$$

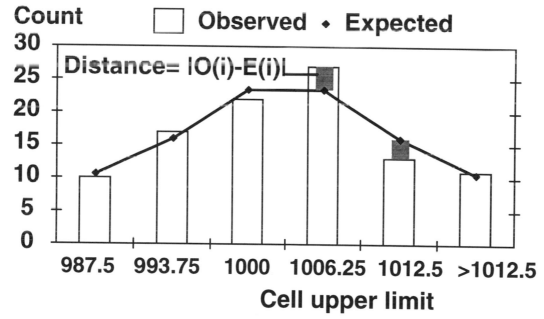

Reject the null hypothesis (normality assumption) if: $\chi^2 > \chi^2_{k-1-p;\alpha}$

Chi Square Test, cont.

- There are $k - 1 - p$ degrees of freedom when we estimate p parameters from the data.

  - If we estimate the mean and variance, then $p = 2$. If we use a standard, then $p = 0$.

  - α is the *significance level* for the test, and is usually 0.05 (5%). It is the risk of wrongly concluding that a normal distribution is not normal.

  - The test will, on average, reject the normality hypothesis 5 times out of 100 when the population is normal.

- For example, 100 measurements come from a population. Test the hypothesis that its mean is 1000 and its variance 100.

Chi Square Test, cont.

Upper limit	Observations	Expected	Calculation
975	1	0.62	$100\Phi\left(\dfrac{975-1000}{10}\right)$
981.25	1	2.42	$100\left(\Phi\left(\dfrac{981.25-1000}{10}\right)-\Phi\left(\dfrac{975-1000}{10}\right)\right)$
987.5	8	7.53	
993.75	17	16.03	
1000	22	23.40	
1006.25	27	23.40	
1012.5	13	16.03	
1018.75	8	7.53	
1025	2	2.42	
>1025	1	0.62	$100\left(1-\Phi\left(\dfrac{1025-1000}{10}\right)\right)$

Transparency Masters to Accompany *SPC Essentials and Productivity Improvement*
ASQC Quality Press   ©1997 by Harris Corporation, Semiconductor Sector

Chi Square Test, cont.

Lower limit	Observation	Expected	χ^2	Sample Calculation
987.5	10	10.56	0.030	$\dfrac{(10-10.56)^2}{10.56}$
993.75	17	16.03	0.058	$\dfrac{(17-16.03)^2}{16.03}$
1000	22	23.40	0.084	
1006.25	27	23.40	0.553	
1012.5	13	16.03	0.574	
≥1012.5	11	10.56	0.018	$\dfrac{(11-10.56)^2}{10.56}$
Total			1.318	

Chi Square Test, cont.

➤ There are six cells.

  ➤ We used standards for μ and σ^2, so $p = 0$. Compare 1.318 against the 95th percentile ($\alpha = 0.05$) χ^2 with 5 degrees of freedom. This is 11.07 > 1.318, so accept the hypothesis that the population is normal.

Transparency Masters to Accompany *SPC Essentials and Productivity Improvement*
ASQC Quality Press   ©1997 by Harris Corporation, Semiconductor Sector

Chi Square Test, cont.

➤ How does this test react when the population is not normal? Consider a bimodal population.

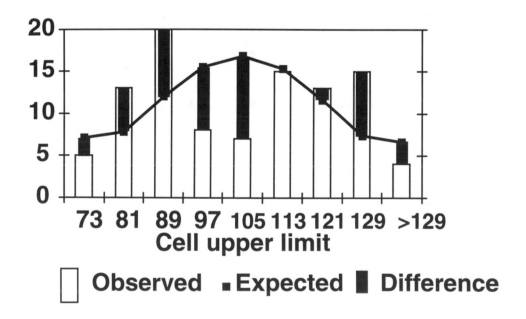

☐ **Observed** ▪ **Expected** ■ **Difference**

Transparency Masters to Accompany *SPC Essentials and Productivity Improvement*
ASQC Quality Press   ©1997 by Harris Corporation, Semiconductor Sector

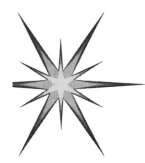

Chi Square Test, cont.

- ➤ After combining cells, there are nine cells.

    - ➤ χ^2 is 28.04.

    - ➤ The mean and standard deviation are estimates from the data, so there are $9 - 3 = 6$ degrees of freedom.

    - ➤ The 95th percentile ($\alpha = 0.05$) for chi square with 6 d.f. is 12.59.

    - ➤ Since 28.04 > 12.59, we are > 95% sure the population is not normal.

Transparency Masters to Accompany *SPC Essentials and Productivity Improvement*
ASQC Quality Press   ©1997 by Harris Corporation, Semiconductor Sector

Normal Probability Plot

➤ The normal probability plot is another useful way to check the normality assumption.

   ➤ It requires less data than the chi square test.

      ➤ Technically, it works with 10 or even fewer data, but less data always reduce a test's power.

➤ Order the data from smallest to largest:

   ➤ $x_{(1)}, x_{(2)}, \ldots, x_{(n)}$. $x_{(i)}$ is the i th ordered or sorted datum.

   ➤ Spreadsheets and MathCAD have functions or tools that do this automatically.

Transparency Masters to Accompany *SPC Essentials and Productivity Improvement*
ASQC Quality Press   ©1997 by Harris Corporation, Semiconductor Sector

Normal Probability Plot, cont.

- A plot of $x_{(i)}$ versus the ordered standard normal deviates $z_{(i)}$ should be linear with slope σ and intercept μ.

- It is not convenient to manually look up a lot of $z_{(i)}$ values.

  - Normal probability paper is available with a percentile scale that is linear in z. To use it, $(i-\frac{1}{2})/n$ for each point must still be calculated.

  - Spreadsheets, however, have built-in functions.

Transparency Masters to Accompany *SPC Essentials and Productivity Improvement*
ASQC Quality Press   ©1997 by Harris Corporation, Semiconductor Sector

Normal Probability Plot, cont.

➤ Example: $n = 50$, $x \sim N(1000, 10^2)$

x (ordered)

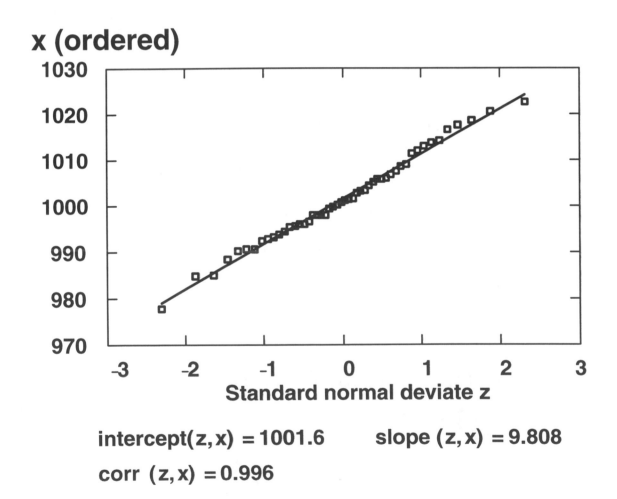

intercept$(z, x) = 1001.6$ slope $(z, x) = 9.808$

corr $(z, x) = 0.996$

Normal Probability Plot, cont.

- ➤ The points fit the line well.
  - ➤ The correlation is good.
  - ➤ The points scatter randomly around the regression line.
  - ➤ Non-normal data will produce poor correlation and systematic patterns around the regression line.
- ➤ The slope is 9.808 versus a sample standard deviation of 9.791.
  - ➤ The slope may match the sample standard deviation, even for non-normal data.
- ➤ Intercept = average = 1001.6.
  - ➤ The intercept will always match the average, even for non-normal data.

Transparency Masters to Accompany *SPC Essentials and Productivity Improvement*
ASQC Quality Press   ©1997 by Harris Corporation, Semiconductor Sector

Normal Probability Plot, cont.

➤ Normal probability plot for bimodal data

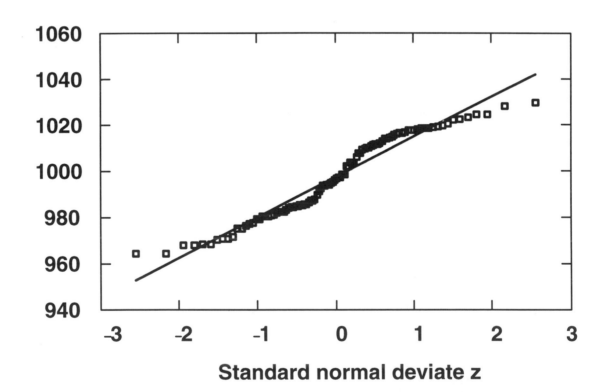

slope (z, x) = 17.419   intercept (z, x) = 997.4

corr (z, x) = 0.976

Normal Probability Plot, cont.

➤ The bimodal data show the following:

- ➤ A systematic pattern around the regression line
- ➤ Poor correlation

➤ *Poor correlation* is a subjective statement.

- ➤ Cryer (1986) gives a table of critical values for correlation in the normal probability plot.

Source: Cryer, Jonathan D. 1986. *Time series analysis*. Boston: PWS-Kent Publishing.

Normal Probability Plot, cont.

➤ Here are the critical values for correlation (Cryer 1986, 43).

	Significance level α		
Sample size	0.10	0.05	0.01
10	0.935	0.918	0.880
15	0.951	0.938	0.911
20	0.960	0.950	0.929
25	0.966	0.958	0.941
30	0.971	0.964	0.949
40	0.977	0.972	0.960
50	0.981	0.976	0.966
60	0.984	0.980	0.971
75	0.987	0.984	0.976
100	0.989	**0.986**	0.981
150	0.992	0.991	0.987
200	0.994	0.993	0.990

Source: Cryer, Jonathan D. 1986. *Time series analysis*. Boston: PWS-Kent Publishing.

Normal Probability Plot, cont.

- ➤ Example: 100 (bimodal) data; correlation = 0.976.

    - ➤ The critical value for 100 data and a 5% significance level is 0.986.

    - ➤ Since 0.976 < 0.986, we are more than 95% sure that the data do not follow the normal distribution.

    - ➤ On average, the correlation of 95 out of 100 sets of 100 normal data will be 0.986 or better.

Transparency Masters to Accompany *SPC Essentials and Productivity Improvement*
ASQC Quality Press   ©1997 by Harris Corporation, Semiconductor Sector

Sample Variance Plot

➤ Given *n* samples, each with *m* measurements

   ➤ Requires constant sample size

Plot the ordered sample variances, $s_{(i)}^2$,

against the $\dfrac{i-0.5}{n}$ percentile of the chi square

distribution with m - 1 degrees of freedom.

The plot should be linear with slope $\dfrac{\sigma^2}{m-1}$

and intercept 0.

Transparency Masters to Accompany *SPC Essentials and Productivity Improvement*
ASQC Quality Press   ©1997 by Harris Corporation, Semiconductor Sector

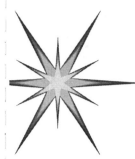

Sample Variance Plot, cont.

➤ This technique requires a computer to be practical.

➤ Spreadsheets have functions for chi square.

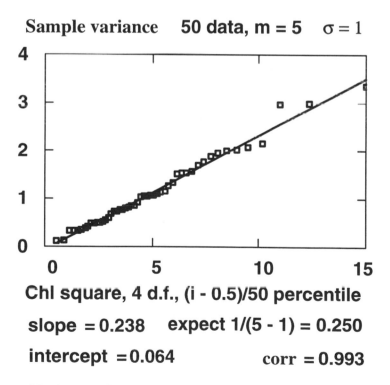

Sample variance   50 data, m = 5 $\sigma = 1$

Chi square, 4 d.f., (i - 0.5)/50 percentile

slope = 0.238   expect 1/(5 - 1) = 0.250

intercept = 0.064   corr = 0.993

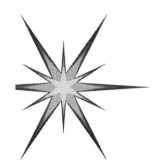

Central Limit Theorem

➤ Averages from large samples follow a normal distribution even when they come from non-normal populations.

    ➤ If the samples are large enough, we can make Shewhart control charts for non-normal populations.

➤ The central limit theorem says that the averages of infinitely large samples follow the normal distribution.

    ➤ In practice, a sample of 5 or 10 may be adequate.

    ➤ It depends on how non-normal the data are.

Transparency Masters to Accompany *SPC Essentials and Productivity Improvement*
ASQC Quality Press   ©1997 by Harris Corporation, Semiconductor Sector

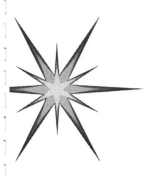

Central Limit Theorem, cont.

➤ Example: 25 samples of 10 from a bimodal population

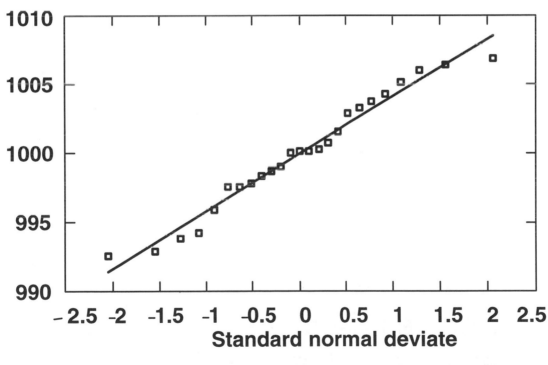

slope (z, x) = 4.165   intercept (z, x) = 1000

corr (z, x) = 0.9873   > 0.958 (5% significance)

Transparency Masters to Accompany *SPC Essentials and Productivity Improvement*
ASQC Quality Press   ©1997 by Harris Corporation, Semiconductor Sector

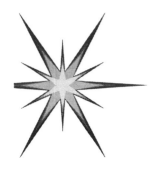

Central Limit Theorem, cont.

- ➤ Alternately, we can characterize the probability distribution.
  - ➤ Fit data to gamma, Weibull, lognormal, and other distributions. Fitting data to some distributions requires advanced techniques.
    - ➤ Lognormal: just see if ln(x) follows a normal distribution.
    - ➤ Gamma: not easy, but some software packages will do it.
  - ➤ Set the LCL at the 0.135 percentile, and the UCL at the 99.835 percentile, to get Shewhart-equivalent false alarm risks.

Transparency Masters to Accompany *SPC Essentials and Productivity Improvement*
ASQC Quality Press   ©1997 by Harris Corporation, Semiconductor Sector

Central Limit Theorem, cont.

- ➤ Applications
  - ➤ The Weibull distribution often applies to lifetime (reliability) data.
  - ➤ The gamma distribution may apply to impurities in chemicals.
    - ➤ A specification can be [0, USL]. Remember, we can't get less than zero!
    - ➤ A one-sided specification is a clue that the distribution may be non-normal.

Transparency Masters to Accompany *SPC Essentials and Productivity Improvement*
ASQC Quality Press   ©1997 by Harris Corporation, Semiconductor Sector

Setting up a Control Chart

- ➤ *Theoretical charts* assume that we know the process mean and standard deviation.

  - ➤ This applies to processes for which we've had long experience.

- ➤ *Empirical charts* rely on estimates for the mean and standard deviation.

Theoretical Control Charts

	Chart for process mean	Chart for process variation	
		s (sample standard deviation) chart	R (sample range) chart
Estimate for standard deviation	$\overline{x}$ chart		
Given	$\mu \pm 3\dfrac{\sigma}{\sqrt{n}}$	Limits $[B_5\sigma, B_6\sigma]$ Centerline: $c_4\sigma$	Limits $[D_1\sigma, D_2\sigma]$ Centerline: $d_2\sigma$

➤ B_5, B_6, D_1, D_2, c_4, and d_2 are factors that depend on sample size.

Transparency Masters to Accompany *SPC Essentials and Productivity Improvement*
ASQC Quality Press   ©1997 by Harris Corporation, Semiconductor Sector

Empirical Control Charts

- Given m samples, each of size n_i: For each sample, we have its standard deviation s_i or range R_i.
  - The standard deviation is better, but it is not practical for manual calculations.
  - We also need the grand average (x-double bar)

Empirical Control Charts, cont.

	Estimate for standard deviation
Based on *s*	$\hat{\sigma} = \dfrac{1}{m}\Sigma^{m}_{i=1}\dfrac{s_i}{c_4(n_i)}$
Based on ranges	$\hat{\sigma} = \dfrac{1}{m}\Sigma^{m}_{i=1}\dfrac{R_i}{d_2(n_i)}$

➤ s/c_4 is an estimate for σ.

➤ R/d_2 is an estimate for σ.

➤ c_4 and d_2 depend on *n.*

Transparency Masters to Accompany *SPC Essentials and Productivity Improvement*
ASQC Quality Press   ©1997 by Harris Corporation, Semiconductor Sector

Empirical Control Charts, cont.

	Chart for process mean	Chart for process variation	
	$\overline{x}$ chart	s chart	R (sample range) chart
Based on s	$\overline{\overline{x}} \pm A_3\overline{s}$	$[B_3\overline{s}, B_4\overline{s}]$ Center line: $\overline{s}$	
Based on ranges	$\overline{\overline{x}} \pm A_2\overline{R}$		$[D_3\overline{R}, D_4\overline{R}]$ Center line: $\overline{R}$

Transparency Masters to Accompany *SPC Essentials and Productivity Improvement*
ASQC Quality Press   ©1997 by Harris Corporation, Semiconductor Sector

Empirical Control Charts, cont.

➤ When the sample sizes vary, so do the control limits and even the centerlines of the *s* and *R* charts.

> ➤ For a sample of *k*, recalling that *s*/*c*$_4$ and *R*/*d*$_2$ are estimates of σ,

$$\overline{s}(k) = c_{4,k}\hat{\sigma} \text{ and } \overline{R}(k) = d_{2,k}\hat{\sigma}$$

Empirical Control Charts, cont.

➤ Calculation for the first sample: $s = 0.5755$, $R = 1.235$. For $n = 5$, $c_4 = 0.9400$, $s/c_4 = 0.6122$.

➤ $d_2 = 2.326$, $R/d_2 = 0.531$.

Sample	n	x-bar	s	s/c_4	R	R/d_2
1	5	100.2	0.576	0.612	1.235	0.531
2	4	100.0	0.336	0.365	0.788	0.383
3	3	100.6	1.132	1.277	2.263	1.337
4	3	99.7	0.769	0.868	1.496	0.884
5	3	100.1	0.359	0.405	0.709	0.419

Transparency Masters to Accompany *SPC Essentials and Productivity Improvement*
ASQC Quality Press   ©1997 by Harris Corporation, Semiconductor Sector

Empirical Control Charts, cont.

Grand average	99.90	s, all data	0.933
$\hat{\sigma}$ based on s values	0.950	on R	0.942

➤ Procedure for the *s* chart

Center line: $\bar{s} = c_{4,n}\,\hat{\sigma} = c_{4,n} \times 0.950$

Control limits: $\left[B_{3,n}\bar{s}, B_{4,n}\bar{s} \right]$

The "n" in the subscript is a reminder that the factor depends on sample size.

Transparency Masters to Accompany *SPC Essentials and Productivity Improvement*
ASQC Quality Press   ©1997 by Harris Corporation, Semiconductor Sector

Empirical Control Charts, cont.

Sample	n	s-bar	s	LCL, s	UCL, s
1	5	0.893	0.576	0.000	1.865
2	4	0.875	0.336	0.000	1.983
3	3	0.842	1.132	0.000	2.162
4	3	0.842	0.769	0.000	2.162
5	3	0.842	0.359	0.000	2.162

➤ Calculation for the first sample

$$\bar{s} = \hat{\sigma} c_4 = 0.950 \times 0.9400 = 0.893$$

$$B_4 \bar{s} = 2.089 \times 0.893 = 1.865 \quad (B_3 = 0)$$

Transparency Masters to Accompany *SPC Essentials and Productivity Improvement*
ASQC Quality Press   ©1997 by Harris Corporation, Semiconductor Sector

Empirical Control Charts, cont.

Samp	n	x-bar	s-bar	$\overline{x}$ chart LCL	$\overline{x}$ chart UCL
1	5	100.2	0.893	98.63	101.17
2	4	100.0	0.875	98.48	101.32
3	3	100.6	0.842	98.25	101.55
4	3	99.7	0.842	98.25	101.55
5	3	100.1	0.842	98.25	101.55

➤ Calculation for the first sample

$$\overline{s} = \hat{\sigma}\, c_4 = 0.950 \times 0.9400 = 0.893$$

$$\overline{\overline{x}} = 99.9 \pm A_3 \overline{s} = 99.9 \pm 1.427 \times 0.893$$

Transparency Masters to Accompany *SPC Essentials and Productivity Improvement*
ASQC Quality Press   ©1997 by Harris Corporation, Semiconductor Sector

Empirical Control
Charts, cont.

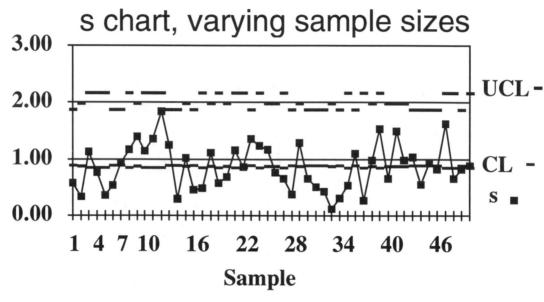

s chart, varying sample sizes

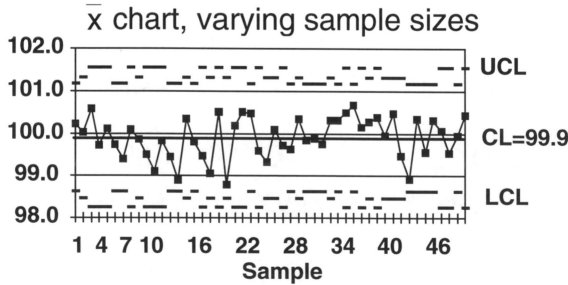

x̄ chart, varying sample sizes

Transparency Masters to Accompany *SPC Essentials and Productivity Improvement*
ASQC Quality Press   ©1997 by Harris Corporation, Semiconductor Sector

Empirical Chart for Individuals (*X* Chart)

➤ In some applications, we can get only samples of one. Use the *X* for process mean.

> ➤ To estimate σ, we can use the average moving range (*MR*).

(1) $MR_i = x_i - x_{i-1}$ for $i = 2$ to m

(2) $\overline{MR} = \dfrac{1}{m-1}\Sigma_{i=2}^{m}|MR_i|$ avg. moving range

| | = absolute value

(3) $\hat{\sigma} = \dfrac{\sqrt{\pi}}{2}\overline{MR}$ $\Rightarrow$

control limits are $\overline{\overline{x}} \pm \dfrac{3\sqrt{\pi}}{2}\overline{MR} = \overline{\overline{x}} \pm 2.66 \times \overline{MR}$

Transparency Masters to Accompany *SPC Essentials and Productivity Improvement*
ASQC Quality Press   ©1997 by Harris Corporation, Semiconductor Sector

X Chart, cont.

➤ Do not make a moving range chart.

  ➤ Unlike a range chart, an *MR* chart does not provide additional information about the process variation.

  ➤ Use the moving ranges only to set the control limits for the *X* chart.

Transparency Masters to Accompany *SPC Essentials and Productivity Improvement*
ASQC Quality Press   ©1997 by Harris Corporation, Semiconductor Sector

X Chart, cont.

➤ Example: 50 samples.

    ➤ Average *MR* = 61.24/49 = 1.25

Sample	X	MR
1	126.02	
2	124.66	−1.36
3	125.89	1.23
4	125.73	−0.16
5	123.52	−2.21
…	…	…
49	124.01	−1.74
50	125.64	1.63
	Σ |*MR*|	61.24

Transparency Masters to Accompany *SPC Essentials and Productivity Improvement*
ASQC Quality Press   ©1997 by Harris Corporation, Semiconductor Sector

X Chart, cont.

$$\hat{\sigma} = \frac{\sqrt{\pi}}{2}1.250 = 1.108, \text{ vs. } s_{DATA} = 1.08$$

Control limits: $124.96 \pm 2.66 \times 1.250$

$= [121.63, \ 128.29]$

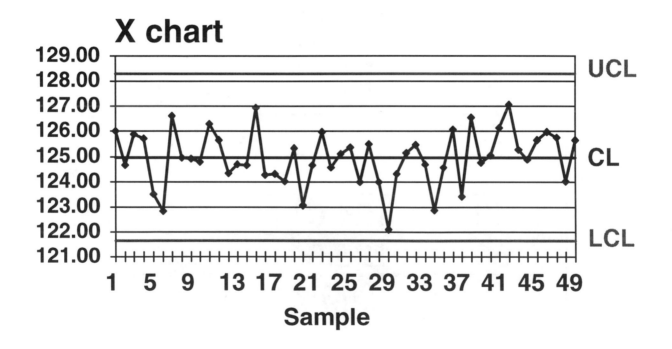

Process Capability Indices

Symbol	Equation	Description
C_p	$\dfrac{USL - LSL}{6\sigma}$	C_p is the ratio of the specification width to the process width. The process width is 6σ.
CPL	$\dfrac{\mu - LSL}{3\sigma}$	CPL measures the process' ability to meet the lower specification.
CPU	$\dfrac{USL - \mu}{3\sigma}$	CPU measures the process' ability to meet the upper specification.
C_{pk}	min [CPL,CPU]	When the process is at the target or nominal, CPL = CPU = C_{pk} = C_p, and the yield is at its maximum.

Transparency Masters to Accompany *SPC Essentials and Productivity Improvement*
ASQC Quality Press   ©1997 by Harris Corporation, Semiconductor Sector

Process Capability Indices, cont.

- ➤ If the process is at its nominal, the specification limits are $3C_p$ standard deviations from the mean.

  - ➤ The nonconforming portions add to $2\Phi(-3C_p)$, so the yield is $1 - 2\Phi(-3C_p)$.

- ➤ If not at nominal, the yield is $1 - \Phi(-3CPL) - \Phi(-3CPU)$.

Process Capability Indices, cont.

➤ Nonconformances vs. C_p (μ = nominal)

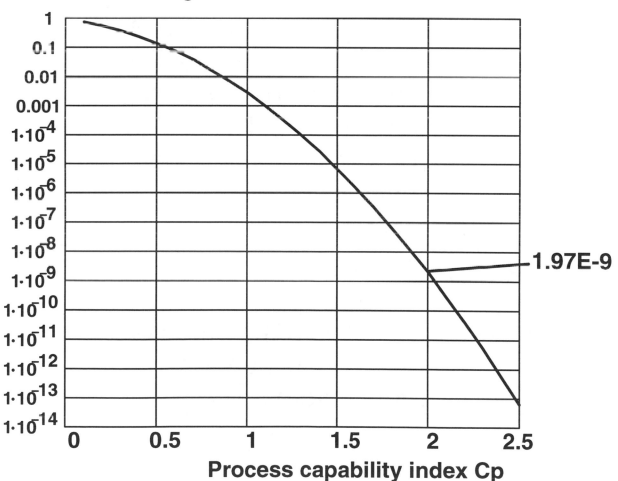

Nonconforming fraction

1.97E-9

Process capability index Cp

Transparency Masters to Accompany *SPC Essentials and Productivity Improvement*
ASQC Quality Press   ©1997 by Harris Corporation, Semiconductor Sector

Process Capability Indices, cont.

➤ ## Capability guidelines

C_p	Process status
$C_p < 1$	Poor. The Shewhart control limits are wider than the specification limits. The process can make bad parts even when it is in control.
$C_p < 1.33$	Fair.
$1.33 \leq C_p$	Acceptable. 1.33 is the basic standard.
$2 \leq C_p$	Excellent. The process will make less than 2 ppb nonconformances.

Process Capability Indices, cont.

- Processes with $C_p < 1.33$ are not capable, and they need improvement.

    - They are muskets instead of rifles.

    - We need to focus on reducing their variation.

- The process mean should be at nominal. This maximizes the yield.

Transparency Masters to Accompany *SPC Essentials and Productivity Improvement*
ASQC Quality Press   ©1997 by Harris Corporation, Semiconductor Sector

Uncertainty in Capability Indices

➤ In practice, we do not know μ or σ. We can only estimate the process capability indices.

$$C_p = \frac{USL - LSL}{6\sigma} \quad \text{vs.} \quad \hat{C}_p = \frac{USL - LSL}{6\hat{\sigma}}$$

$$\text{Also,} \quad \hat{C}PL = \frac{\hat{\mu} - LSL}{3\hat{\sigma}} \quad \text{and} \quad \hat{C}PU = \frac{USL - \hat{\mu}}{3\hat{\sigma}}$$

$100(1 - \alpha)\%$ confidence interval for C_p

$$\hat{C}_p \sqrt{\frac{\chi^2_{1-\alpha/2;n-1}}{n-1}} \leq C_p \leq \hat{C}_p \sqrt{\frac{\chi^2_{\alpha/2;n-1}}{n-1}}$$

where $\chi^2_{\alpha/2;n-1}$ is the $100\left(1 - \frac{\alpha}{2}\right)$ percentile for χ^2

with n - 1 degrees of freedom

# Uncertainty, cont.

➤ 95% (two-sided) confidence interval for C_p.

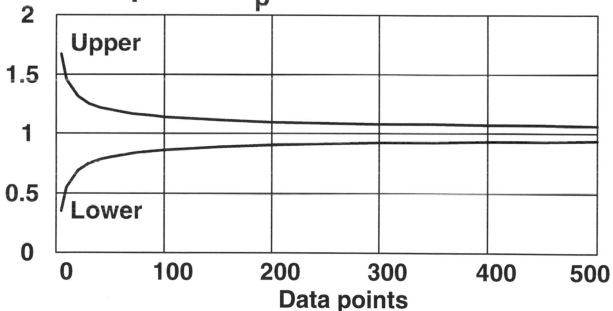

Multiple of $\hat{C}_p$

95% (2 - sided) confidence limits for C_p,

given 100 data: $[0.86\hat{C}_p, 1.14\hat{C}_p]$

Special Control Charts

➤ The *z* chart is for equipment that operates on different products and whose settings change frequently.

➤ The *center band chart* or *acceptance control chart* is for processes with an acceptable range of means.

Transparency Masters to Accompany *SPC Essentials and Productivity Improvement*
ASQC Quality Press   ©1997 by Harris Corporation, Semiconductor Sector

The z Chart

➤ The z chart is useful in job-shop operations where a tool may process different products.

  ➤ Each product may have its own specifications, nominal, and standard deviation.

  ➤ It is inconvenient to keep a separate chart for each product.

  ➤ Separate charts cannot efficiently use the Western Electric zone tests.

The *z* Chart, cont.

- ➤ To plot the average of *n* measurements for product type *j*, calculate *z* as follows.
  - ➤ μ_j = nominal for type *j*
  - ➤ σ_j = standard deviation for type *j*
  - ➤ LCL = −3, UCL = 3

$$z = \frac{\overline{x} - \mu_j}{\dfrac{\sigma_j}{\sqrt{n}}}$$

The *z* Chart, cont.: The χ^2 Chart

➤ How can we track variation for several products on a common chart?

Test the hypothesis $\sigma^2 = \sigma_0^2$

For a sample of n, jth type: $\chi^2 = \dfrac{(n-1)s^2}{\sigma_{0,j}^2}$

Control limits: $\left[\dfrac{\alpha}{2}, \ 1-\dfrac{\alpha}{2}\right]$ quantiles for χ^2

with n - 1 degrees of freedom

The z and χ^2 Chart, Example

Sample size = 5

i	μ_0	σ_0	x_bar	z
1	300	3	300.07	0.05
2	300	3	297.52	−1.85
3	300	3	299.97	−0.03
4	100	1	100.31	0.69
…	…	…	…	…
9	100	1	99.51	−1.09
10	200	2	201.05	1.17

i	σ_0	s	χ^2	
1	3	0.869	0.34	LCL = 0.106
2	3	2.589	2.98	CL = 3.36
3	3	3.716	6.14	UCL = 17.80
4	1	0.958	3.67	
…	…	…	…	
9	1	1.102	4.86	
10	2	1.832	3.36	

Transparency Masters to Accompany *SPC Essentials and Productivity Improvement*
ASQC Quality Press   ©1997 by Harris Corporation, Semiconductor Sector

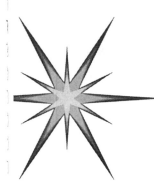

z and Chi Square Charts

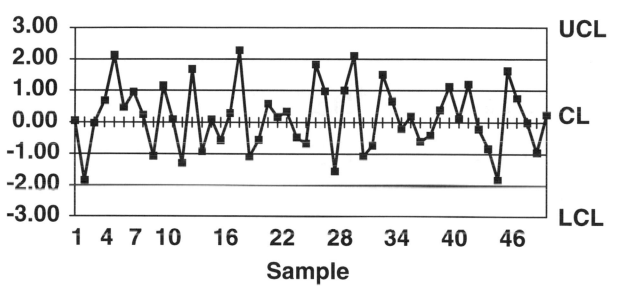

χ^2 Centerline = median, with 4 d.f.

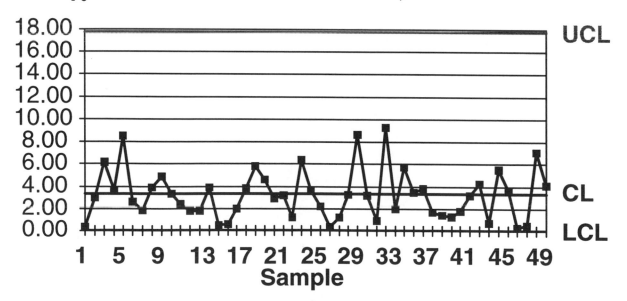

Transparency Masters to Accompany *SPC Essentials and Productivity Improvement*
ASQC Quality Press   ©1997 by Harris Corporation, Semiconductor Sector

Charts with Center Bands

> The center band chart is for processes that have acceptable ranges for their means.

> > It may be impractical to set the process to a single mean.

> > The process mean is acceptable within [LAPL, UAPL].

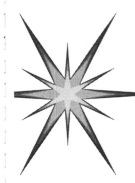

Charts with Center Bands, cont.

➤ Example: LAPL and UAPL are the lower and upper acceptable process limits.

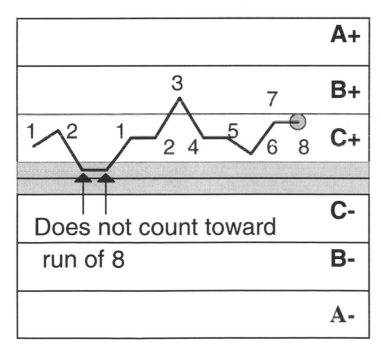

A+	UCL = UAPL +3σ
	UAPL + 2σ
B+	UAPL + σ
C+	UAPL
	LAPL
C-	LAPL -σ
B-	LAPL - 2σ
A-	LCL = LAPL -3σ

Does not count toward run of 8

A+, A-, etc. = zones for Western Electric tests

Transparency Masters to Accompany *SPC Essentials and Productivity Improvement*
ASQC Quality Press   ©1997 by Harris Corporation, Semiconductor Sector

Charts with Center Bands, cont.

➤ The book gives a procedure for customizing control charts to meet specific requirements.

  ➤ β is the risk of accepting the process when the mean is outside the rejectable process limits.

  ➤ α is the chance of rejecting it when it is within [LAPL, UAPL].

False Alarm Risks and Power

- ➤ A chart's *power* is its ability to detect an undesirable shift.
  - ➤ It depends on the control limits, sample size, and the shift.
- ➤ Consider a process shift of $\delta\sigma$.
  - ➤ The UCL is $\mu + 3\sigma/n^{0.5}$.
  - ➤ What is the chance of detecting it?

$$Pr(\bar{x} \leq UCL) = \Phi\left(\frac{\left(\mu + \dfrac{3\sigma}{\sqrt{n}}\right) - (\mu + \delta\sigma)}{\dfrac{\sigma}{\sqrt{n}}}\right)$$

$$= \Phi\left(3 - \delta\sqrt{n}\right)$$

$$\text{Power } \gamma(\delta, n) = 1 - \Phi\left(3 - \delta\sqrt{n}\right) = Pr(\bar{x} \geq UCL)$$

Transparency Masters to Accompany *SPC Essentials and Productivity Improvement*
ASQC Quality Press   ©1997 by Harris Corporation, Semiconductor Sector

False Alarm Risks and Power, cont.

➤ Control chart power versus process shift

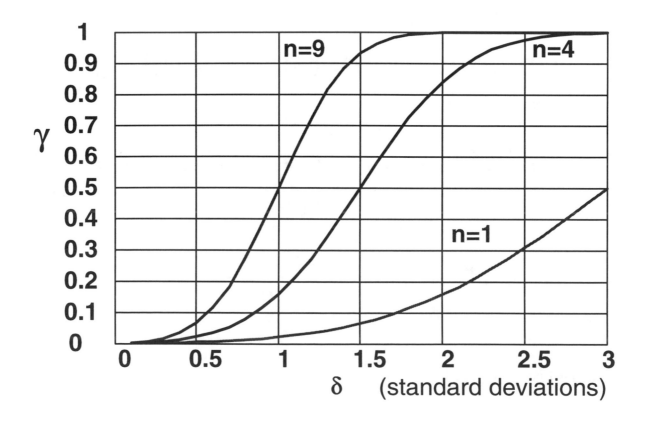

Transparency Masters to Accompany *SPC Essentials and Productivity Improvement*
ASQC Quality Press   ©1997 by Harris Corporation, Semiconductor Sector

Average Run Length

➤ The reciprocal of the power is the *average run length* (ARL).

  ➤ It is the number of samples we expect to measure before detecting a process shift.

  ➤ When there is no process shift, the false alarm risk is always 2.7 per thousand samples. We can increase a chart's power by increasing the sample size.

➤ We can also calculate ARLs and powers for the Western Electric zone tests.

  ➤ The book goes into the details.

Transparency Masters to Accompany *SPC Essentials and Productivity Improvement*
ASQC Quality Press   ©1997 by Harris Corporation, Semiconductor Sector

Average Run Length, cont.

➤ ARL (one-sided test, UCL or LCL)

Average run length (log scale)

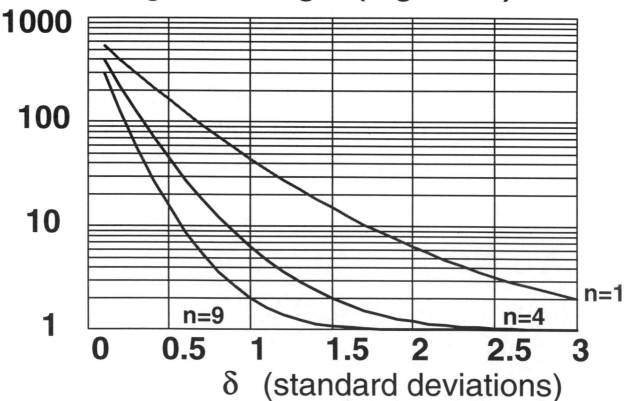

Transparency Masters to Accompany *SPC Essentials and Productivity Improvement*
ASQC Quality Press   ©1997 by Harris Corporation, Semiconductor Sector

APPENDIX A

Percentiles of the Standard Normal Distribution*

$$\Phi(z) = \int_{-\infty}^{z} \frac{1}{\sqrt{2\pi}} \exp\left(-\frac{x^2}{2}\right) dx$$

The left column shows the ones and tenths places of the standard normal deviate. The remaining columns show the hundredths. For example, the standard normal deviate of 1.96 is $\Phi(1.96) = 0.975$.

---

*Using Microsoft Excel's NORMSDIST function.

	0.00	0.01	0.02	0.03	0.04	0.05	0.06	0.07	0.08	0.09
0.0	0.5000	0.5040	0.5080	0.5120	0.5160	0.5199	0.5239	0.5279	0.5319	0.5359
0.1	0.5398	0.5438	0.5478	0.5517	0.5557	0.5596	0.5636	0.5675	0.5714	0.5753
0.2	0.5793	0.5832	0.5871	0.5910	0.5948	0.5987	0.6026	0.6064	0.6103	0.6141
0.3	0.6179	0.6217	0.6255	0.6293	0.6331	0.6368	0.6406	0.6443	0.6480	0.6517
0.4	0.6554	0.6591	0.6628	0.6664	0.6700	0.6736	0.6772	0.6808	0.6844	0.6879
0.5	0.6915	0.6950	0.6985	0.7019	0.7054	0.7088	0.7123	0.7157	0.7190	0.7224
0.6	0.7257	0.7291	0.7324	0.7357	0.7389	0.7422	0.7454	0.7486	0.7517	0.7549
0.7	0.7580	0.7611	0.7642	0.7673	0.7704	0.7734	0.7764	0.7794	0.7823	0.7852
0.8	0.7881	0.7910	0.7939	0.7967	0.7995	0.8023	0.8051	0.8078	0.8106	0.8133
0.9	0.8159	0.8186	0.8212	0.8238	0.8264	0.8289	0.8315	0.8340	0.8365	0.8389
1.0	0.8413	0.8438	0.8461	0.8485	0.8508	0.8531	0.8554	0.8577	0.8599	0.8621
1.1	0.8643	0.8665	0.8686	0.8708	0.8729	0.8749	0.8770	0.8790	0.8810	0.8830
1.2	0.8849	0.8869	0.8888	0.8907	0.8925	0.8944	0.8962	0.8980	0.8997	0.90147
1.3	0.90320	0.90490	0.90658	0.90824	0.90988	0.91149	0.91308	0.91466	0.91621	0.91774
1.4	0.91924	0.92073	0.92220	0.92364	0.92507	0.92647	0.92785	0.92922	0.93056	0.93189
1.5	0.93319	0.93448	0.93574	0.93699	0.93822	0.93943	0.94062	0.94179	0.94295	0.94408
1.6	0.94520	0.94630	0.94738	0.94845	0.94950	0.95053	0.95154	0.95254	0.95352	0.95449
1.7	0.95543	0.95637	0.95728	0.95818	0.95907	0.95994	0.96080	0.96164	0.96246	0.96327
1.8	0.96407	0.96485	0.96562	0.96638	0.96712	0.96784	0.96856	0.96926	0.96995	0.97062
1.9	0.97128	0.97193	0.97257	0.97320	0.97381	0.97441	0.97500	0.97558	0.97615	0.97670
2.0	0.97725	0.97778	0.97831	0.97882	0.97932	0.97982	0.98030	0.98077	0.98124	0.98169
2.1	0.98214	0.98257	0.98300	0.98341	0.98382	0.98422	0.98461	0.98500	0.98537	0.98574
2.2	0.98610	0.98645	0.98679	0.98713	0.98745	0.98778	0.98809	0.98840	0.98870	0.98899
2.3	0.98928	0.98956	0.98983	0.990097	0.990358	0.990613	0.990863	0.991106	0.991344	0.991576
2.4	0.991802	0.992024	0.992240	0.992451	0.992656	0.992857	0.993053	0.993244	0.993431	0.993613
2.5	0.993790	0.993963	0.994132	0.994297	0.994457	0.994614	0.994766	0.994915	0.995060	0.995201
2.6	0.995339	0.995473	0.995603	0.995731	0.995855	0.995975	0.996093	0.996207	0.996319	0.996427
2.7	0.996533	0.996636	0.996736	0.996833	0.996928	0.997020	0.997110	0.997197	0.997282	0.997365
2.8	0.997445	0.997523	0.997599	0.997673	0.997744	0.997814	0.997882	0.997948	0.998012	0.998074
2.9	0.998134	0.998193	0.998250	0.998305	0.998359	0.998411	0.998462	0.998511	0.998559	0.998605
3.0	0.998650	0.998694	0.998736	0.998777	0.998817	0.998856	0.998893	0.998930	0.998965	0.998999
3.1	0.9990323	0.9990645	0.9990957	0.9991259	0.9991552	0.9991836	0.9992111	0.9992377	0.9992636	0.9992886
3.2	0.9993128	0.9993363	0.9993590	0.9993810	0.9994023	0.9994229	0.9994429	0.9994622	0.9994809	0.9994990
3.3	0.9995165	0.9995335	0.9995499	0.9995657	0.9995811	0.9995959	0.9996102	0.9996241	0.9996375	0.9996505
3.4	0.9996630	0.9996751	0.9996868	0.9996982	0.9997091	0.9997197	0.9997299	0.9997397	0.9997492	0.9997584
3.5	0.9997673	0.9997759	0.9997842	0.9997922	0.9997999	0.9998073	0.9998145	0.9998215	0.9998282	0.9998346
3.6	0.9998409	0.9998469	0.9998527	0.9998583	0.9998636	0.9998688	0.9998739	0.9998787	0.9998834	0.9998878
3.7	0.9998922	0.9998963	0.99990036	0.99990423	0.99990796	0.99991156	0.99991502	0.99991835	0.99992156	0.99992465
3.8	0.99992763	0.99993049	0.99993325	0.99993591	0.99993846	0.99994092	0.99994329	0.99994556	0.99994775	0.99994986
3.9	0.99995188	0.99995383	0.99995571	0.99995751	0.99995924	0.99996091	0.99996251	0.99996405	0.99996553	0.99996695
4.0	0.99996831	0.99996963	0.99997089	0.99997210	0.99997326	0.99997438	0.99997545	0.99997648	0.99997747	0.99997842

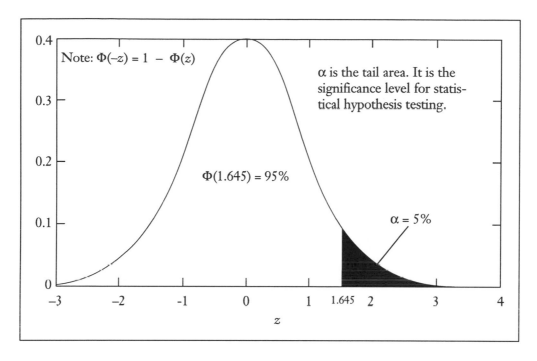

Figure A.1. Percentiles of the standard normal distribution.

Percentiles of the Chi Square Distribution*

$$F(\chi^2) = \int_0^{\chi^2} \frac{1}{2^{n/2}\Gamma\left(\dfrac{n}{2}\right)}\; x^{\frac{n-2}{2}} \exp\left(-\frac{x}{2}\right) dx \quad \text{for } n \text{ degrees of freedom}$$

*Using Microsoft Excel's CHIINV function.

	Cumulative chi square (F)												
n	0.005	0.01	0.025	0.05	0.1	0.25	0.5	0.75	0.9	0.95	0.975	0.99	0.995
1	3.93E – 05	1.57E – 04	9.82E – 04	3.93E – 03	1.58E – 02	0.102	0.455	1.32	2.71	3.84	5.02	6.63	7.88
2	1.00E – 02	2.01E – 02	5.06E – 02	0.103	0.211	10.575	1.39	2.77	4.61	5.99	7.38	9.21	10.60
3	7.17E – 02	0.115	0.216	0.352	0.584	1.21	2.37	4.11	6.25	7.81	9.35	11.34	12.84
4	0.21	0.30	0.48	0.71	1.06	1.92	3.36	5.39	7.78	9.49	11.14	13.28	14.86
5	0.41	0.55	0.83	1.15	1.61	2.67	4.35	6.63	9.24	11.07	12.83	15.09	16.75
6	0.68	0.87	1.24	1.64	2.20	3.45	5.35	7.84	10.64	12.59	14.45	16.81	18.55
7	0.99	1.24	1.69	2.17	2.83	4.25	6.35	9.04	12.02	14.07	16.01	18.48	20.28
8	1.34	1.65	2.18	2.73	3.49	5.07	7.34	10.22	13.36	15.51	17.53	20.09	21.95
9	1.73	2.09	2.70	3.33	4.17	5.90	8.34	11.39	14.68	16.92	19.02	21.67	23.59
10	2.16	2.56	3.25	3.94	4.87	6.74	9.34	12.55	15.99	18.31	20.48	23.21	25.19
11	2.60	3.05	3.82	4.57	5.58	7.58	10.34	13.70	17.28	19.68	21.92	24.73	26.76
12	3.07	3.57	4.40	5.23	6.30	8.44	11.34	14.85	18.55	21.03	23.34	26.22	28.30
13	3.57	4.11	5.01	5.89	7.04	9.30	12.34	15.98	19.81	22.36	24.74	27.69	29.82
14	4.07	4.66	5.63	6.57	7.79	10.17	13.34	17.12	21.06	23.68	26.12	29.14	31.32
15	4.60	5.23	6.26	7.26	8.55	11.04	14.34	18.25	22.31	25.00	27.49	30.58	32.80
16	5.14	5.81	6.91	7.96	9.31	11.91	15.34	19.37	23.54	26.30	28.85	32.00	34.27
17	5.70	6.41	7.56	8.67	10.09	12.79	16.34	20.49	24.77	27.59	30.19	33.41	35.72
18	6.26	7.01	8.23	9.39	10.86	13.68	17.34	21.60	25.99	28.87	31.53	34.81	37.16
19	6.84	7.63	8.91	10.12	11.65	14.56	18.34	22.72	27.20	30.14	32.85	36.19	38.58
20	7.43	8.26	9.59	10.85	12.44	15.45	19.34	23.83	28.41	31.41	34.17	37.57	40.00
21	8.03	8.90	10.28	11.59	13.24	16.34	20.34	24.93	29.62	32.67	35.48	38.93	41.40
22	8.64	9.54	10.98	12.34	14.04	17.24	21.34	26.04	30.81	33.92	36.78	40.29	42.80
23	9.26	10.20	11.69	13.09	14.85	18.14	22.34	27.14	32.01	35.17	38.08	41.64	44.18
24	9.89	10.86	12.40	13.85	15.66	19.04	23.34	28.24	33.20	36.42	39.36	42.98	45.56
25	10.52	11.52	13.12	14.61	16.47	19.94	24.34	29.34	34.38	37.65	40.65	44.31	46.93
26	11.16	12.20	13.84	15.38	17.29	20.84	25.34	30.43	35.56	38.89	41.92	45.64	48.29
27	11.81	12.88	14.57	16.15	18.11	21.75	26.34	31.53	36.74	40.11	43.19	46.96	49.65
28	12.46	13.56	15.31	16.93	18.94	22.66	27.34	32.62	37.92	41.34	44.46	48.28	50.99
29	13.12	14.26	16.05	17.71	19.77	23.57	28.34	33.71	39.09	42.56	45.72	49.59	52.34
30	13.79	14.95	16.79	18.49	20.60	24.48	29.34	34.80	40.26	43.77	46.98	50.89	53.67
31	14.46	15.66	17.54	19.28	21.43	25.39	30.34	35.89	41.42	44.99	48.23	52.19	55.00
32	15.13	16.36	18.29	20.07	22.27	26.30	31.34	36.97	42.58	46.19	49.48	53.49	56.33
33	15.82	17.07	19.05	20.87	23.11	27.22	32.34	38.06	43.75	47.40	50.73	54.78	57.65
34	16.50	17.79	19.81	21.66	23.95	28.14	33.34	39.14	44.90	48.60	51.97	56.06	58.96
35	17.19	18.51	20.57	22.47	24.80	29.05	34.34	40.22	46.06	49.80	53.20	57.34	60.27
36	17.89	19.23	21.34	23.27	25.64	29.97	35.34	41.30	47.21	51.00	54.44	58.62	61.58
37	18.59	19.96	22.11	24.07	26.49	30.89	36.34	42.38	48.36	52.19	55.67	59.89	62.88
38	19.29	20.69	22.88	24.88	27.34	31.81	37.34	43.46	49.51	53.38	56.90	61.16	64.18
39	20.00	21.43	23.65	25.70	28.20	32.74	38.34	44.54	50.66	54.57	58.12	62.43	65.48
40	20.71	22.16	24.43	26.51	29.05	33.66	39.34	45.62	51.81	55.76	59.34	63.69	66.77
41	21.42	22.91	25.21	27.33	29.91	34.58	40.34	46.69	52.95	56.94	60.56	64.95	68.05
42	22.14	23.65	26.00	28.14	30.77	35.51	41.34	47.77	54.09	58.12	61.78	66.21	69.34
43	22.86	24.40	26.79	28.96	31.63	36.44	42.34	48.84	55.23	59.30	62.99	67.46	70.62
44	23.58	25.15	27.57	29.79	32.49	37.36	43.34	49.91	56.37	60.48	64.20	68.71	71.89
45	24.31	25.90	28.37	30.61	33.35	38.29	44.34	50.98	57.51	61.66	65.41	69.96	73.17
46	25.04	26.66	29.16	31.44	34.22	39.22	45.34	52.06	58.64	62.83	66.62	71.20	74.44
47	25.77	27.42	29.96	32.27	35.08	40.15	46.34	53.13	59.77	64.00	67.82	72.44	75.70
48	26.51	28.18	30.75	33.10	35.95	41.08	47.34	54.20	60.91	65.17	69.02	73.68	76.97
49	27.25	28.94	31.55	33.93	36.82	42.01	48.33	55.27	62.04	66.34	70.22	74.92	78.23
50	27.99	29.71	32.36	34.76	37.69	42.94	49.33	56.33	63.17	67.50	71.42	76.15	79.49

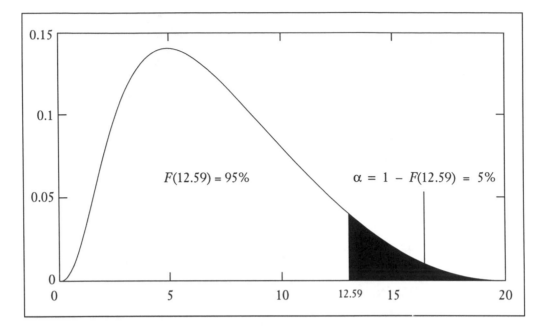

Figure B.1. Chi square distribution (6 degrees of freedom).

For $n > 30$, χ^2_p; $n \approx \frac{1}{2}\left[z_p + \sqrt{2n-1}\right]^2$ where z_p is the pth percentile of the standard normal distribution. For example, $z_{0.05} = -1.645$ and $\frac{1}{2}\left[-1.645 + \sqrt{2(50-1)}\right]^2 = 34.49$, while the fifth percentile for χ^2 with 50 degrees of freedom is 34.76. The approximation improves as n increases.

Percentiles of the Student's *t* Distribution*

$$F(t) = \int_{-\infty}^{z} \frac{\Gamma\left(\frac{n+1}{2}\right)}{\sqrt{n\pi}\ \Gamma\left(\frac{n}{2}\right)} \left(1 + \frac{x^2}{n}\right)^{-\frac{n+1}{2}} dx \quad \text{for } n \text{ degrees of freedom}$$

---

*Using Microsoft Excel's TINV function.

	Cumulative t distribution $F(t)$						
n	0.75	0.9	0.95	0.975	0.99	0.995	0.999
1	1.000	3.078	6.314	12.706	31.821	63.656	318.289
2	0.816	1.886	2.920	4.303	6.965	9.925	22.328
3	0.765	1.638	2.353	3.182	4.541	5.841	10.214
4	0.741	1.533	2.132	2.776	3.747	4.604	7.173
5	0.727	1.476	2.015	2.571	3.365	4.032	5.894
6	0.718	1.440	1.943	2.447	3.143	3.707	5.208
7	0.711	1.415	1.895	2.365	2.998	3.499	4.785
8	0.706	1.397	1.860	2.306	2.896	3.355	4.501
9	0.703	1.383	1.833	2.262	2.821	3.250	4.297
10	0.700	1.372	1.812	2.228	2.764	3.169	4.144
11	0.697	1.363	1.796	2.201	2.718	3.106	4.025
12	0.695	1.356	1.782	2.179	2.681	3.055	3.930
13	0.694	1.350	1.771	2.160	2.650	3.012	3.852
14	0.692	1.345	1.761	2.145	2.624	2.977	3.787
15	0.691	1.341	1.753	2.131	2.602	2.947	3.733
16	0.690	1.337	1.746	2.120	2.583	2.921	3.686
17	0.689	1.333	1.740	2.110	2.567	2.898	3.646
18	0.688	1.330	1.734	2.101	2.552	2.878	3.610
19	0.688	1.328	1.729	2.093	2.539	2.861	3.579
20	0.687	1.325	1.725	2.086	2.528	2.845	3.552
21	0.686	1.323	1.721	2.080	2.518	2.831	3.527
22	0.686	1.321	1.717	2.074	2.508	2.819	3.505
23	0.685	1.319	1.714	2.069	2.500	2.807	3.485
24	0.685	1.318	1.711	2.064	2.492	2.797	3.467
25	0.684	1.316	1.708	2.060	2.485	2.787	3.450
26	0.684	1.315	1.706	2.056	2.479	2.779	3.435
27	0.684	1.314	1.703	2.052	2.473	2.771	3.421
28	0.683	1.313	1.701	2.048	2.467	2.763	3.408
29	0.683	1.311	1.699	2.045	2.462	2.756	3.396
30	0.683	1.310	1.697	2.042	2.457	2.750	3.385
40	0.681	1.303	1.684	2.021	2.423	2.704	3.307
50	0.679	1.299	1.676	2.009	2.403	2.678	3.261
60	0.679	1.296	1.671	2.000	2.390	2.660	3.232
70	0.678	1.294	1.667	1.994	2.381	2.648	3.211
80	0.678	1.292	1.664	1.990	2.374	2.639	3.195
90	0.677	1.291	1.662	1.987	2.368	2.632	3.183
100	0.677	1.290	1.660	1.984	2.364	2.626	3.174
110	0.677	1.289	1.659	1.982	2.361	2.621	3.166
120	0.677	1.289	1.658	1.980	2.358	2.617	3.160
1000000	0.674	1.282	1.645	1.960	2.326	2.576	3.090

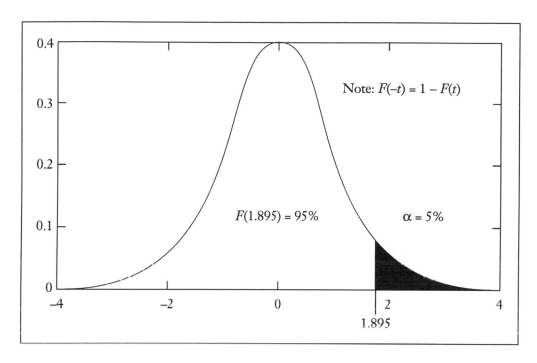

Figure C.1. *t* distribution (7 degrees of freedom).

APPENDIX D

F **Distribution Tables***

$$F(F) = \int_0^F \frac{\Gamma\left(\dfrac{m+n}{2}\right)}{\Gamma\left(\dfrac{m}{2}\right)\Gamma\left(\dfrac{n}{2}\right)} m^{\frac{m}{2}} n^{\frac{n}{2}} x^{\frac{m}{2}-1} (n + mx)^{-\frac{m+n}{2}} dx$$

for *m* degrees of freedom in the numerator, and *n* in the denominator

---

*Using Microsoft Excel's FINV function.

Table D.1. 90th percentile of the *F* distribution.

n	Numerator (m)																	
	1	2	3	4	5	6	7	8	9	10	12	15	20	24	30	40	60	120
1	39.86	49.50	53.59	55.83	57.24	58.20	58.91	59.44	59.86	60.19	60.71	61.22	61.74	62.00	62.26	62.53	62.79	63.06
2	8.53	9.00	9.16	9.24	9.29	9.33	9.35	9.37	9.38	9.39	9.41	9.42	9.44	9.45	9.46	9.47	9.47	9.48
3	5.54	5.46	5.39	5.34	5.31	5.28	5.27	5.25	5.24	5.23	5.22	5.20	5.18	5.18	5.17	5.16	5.15	5.14
4	4.54	4.32	4.19	4.11	4.05	4.01	3.98	3.95	3.94	3.92	3.90	3.87	3.84	3.83	3.82	3.80	3.79	3.78
5	4.06	3.78	3.62	3.52	3.45	3.40	3.37	3.34	3.32	3.30	3.27	3.24	3.21	3.19	3.17	3.16	3.14	3.12
6	3.78	3.46	3.29	3.18	3.11	3.05	3.01	2.98	2.96	2.94	2.90	2.87	2.84	2.82	2.80	2.78	2.76	2.74
7	3.59	3.26	3.07	2.96	2.88	2.83	2.78	2.75	2.72	2.70	2.67	2.63	2.59	2.58	2.56	2.54	2.51	2.49
8	3.46	3.11	2.92	2.81	2.73	2.67	2.62	2.59	2.56	2.54	2.50	2.46	2.42	2.40	2.38	2.36	2.34	2.32
9	3.36	3.01	2.81	2.69	2.61	2.55	2.51	2.47	2.44	2.42	2.38	2.34	2.30	2.28	2.25	2.23	2.21	2.18
10	3.29	2.92	2.73	2.61	2.52	2.46	2.41	2.38	2.35	2.32	2.28	2.24	2.20	2.18	2.16	2.13	2.11	2.08
11	3.23	2.86	2.66	2.54	2.45	2.39	2.34	2.30	2.27	2.25	2.21	2.17	2.12	2.10	2.08	2.05	2.03	2.00
12	3.18	2.81	2.61	2.48	2.39	2.33	2.28	2.24	2.21	2.19	2.15	2.10	2.06	2.04	2.01	1.99	1.96	1.93
13	3.14	2.76	2.56	2.43	2.35	2.28	2.23	2.20	2.16	2.14	2.10	2.05	2.01	1.98	1.96	1.93	1.90	1.88
14	3.10	2.73	2.52	2.39	2.31	2.24	2.19	2.15	2.12	2.10	2.05	2.01	1.96	1.94	1.91	1.89	1.86	1.83
15	3.07	2.70	2.49	2.36	2.27	2.21	2.16	2.12	2.09	2.06	2.02	1.97	1.92	1.90	1.87	1.85	1.82	1.79
16	3.05	2.67	2.46	2.33	2.24	2.18	2.13	2.09	2.06	2.03	1.99	1.94	1.89	1.87	1.84	1.81	1.78	1.75
17	3.03	2.64	2.44	2.31	2.22	2.15	2.10	2.06	2.03	2.00	1.96	1.91	1.86	1.84	1.81	1.78	1.75	1.72
18	3.01	2.62	2.42	2.29	2.20	2.13	2.08	2.04	2.00	1.98	1.93	1.89	1.84	1.81	1.78	1.75	1.72	1.69
19	2.99	2.61	2.40	2.27	2.18	2.11	2.06	2.02	1.98	1.96	1.91	1.86	1.81	1.79	1.76	1.73	1.70	1.67
20	2.97	2.59	2.38	2.25	2.16	2.09	2.04	2.00	1.96	1.94	1.89	1.84	1.79	1.77	1.74	1.71	1.68	1.64
21	2.96	2.57	2.36	2.23	2.14	2.08	2.02	1.98	1.95	1.92	1.87	1.83	1.78	1.75	1.72	1.69	1.66	1.62
22	2.95	2.56	2.35	2.22	2.13	2.06	2.01	1.97	1.93	1.90	1.86	1.81	1.76	1.73	1.70	1.67	1.64	1.60
23	2.94	2.55	2.34	2.21	2.11	2.05	1.99	1.95	1.92	1.89	1.84	1.80	1.74	1.72	1.69	1.66	1.62	1.59
24	2.93	2.54	2.33	2.19	2.10	2.04	1.98	1.94	1.91	1.88	1.83	1.78	1.73	1.70	1.67	1.64	1.61	1.57
25	2.92	2.53	2.32	2.18	2.09	2.02	1.97	1.93	1.89	1.87	1.82	1.77	1.72	1.69	1.66	1.63	1.59	1.56
26	2.91	2.52	2.31	2.17	2.08	2.01	1.96	1.92	1.88	1.86	1.81	1.76	1.71	1.68	1.65	1.61	1.58	1.54
27	2.90	2.51	2.30	2.17	2.07	2.00	1.95	1.91	1.87	1.85	1.80	1.75	1.70	1.67	1.64	1.60	1.57	1.53
28	2.89	2.50	2.29	2.16	2.06	2.00	1.94	1.90	1.87	1.84	1.79	1.74	1.69	1.66	1.63	1.59	1.56	1.52
29	2.89	2.50	2.28	2.15	2.06	1.99	1.93	1.89	1.86	1.83	1.78	1.73	1.68	1.65	1.62	1.58	1.55	1.51
30	2.88	2.49	2.28	2.14	2.05	1.98	1.93	1.88	1.85	1.82	1.77	1.72	1.67	1.64	1.61	1.57	1.54	1.50
40	2.84	2.44	2.23	2.09	2.00	1.93	1.87	1.83	1.79	1.76	1.71	1.66	1.61	1.57	1.54	1.51	1.47	1.42
60	2.79	2.39	2.18	2.04	1.95	1.87	1.82	1.77	1.74	1.71	1.66	1.60	1.54	1.51	1.48	1.44	1.40	1.35
120	2.75	2.35	2.13	1.99	1.90	1.82	1.77	1.72	1.68	1.65	1.60	1.55	1.48	1.45	1.41	1.37	1.32	1.26

Table D.2. 95th percentile of the F distribution.

n	\multicolumn{18}{c}{Numerator (m)}
	1	2	3	4	5	6	7	8	9	10	12	15	20	24	30	40	60	120
1	161.4	199.5	215.7	224.6	230.2	234.0	236.8	238.9	240.5	241.9	243.9	245.9	248.0	249.1	250.1	251.1	252.2	253.3
2	18.51	19.00	19.16	19.25	19.30	19.33	19.35	19.37	19.38	19.40	19.41	19.43	19.45	19.45	19.46	19.47	19.48	19.49
3	10.13	9.55	9.28	9.12	9.01	8.94	8.89	8.85	8.81	8.79	8.74	8.70	8.66	8.64	8.62	8.59	8.57	8.55
4	7.71	6.94	6.59	6.39	6.26	6.16	6.09	6.04	6.00	5.96	5.91	5.86	5.80	5.77	5.75	5.72	5.69	5.66
5	6.61	5.79	5.41	5.19	5.05	4.95	4.88	4.82	4.77	4.74	4.68	4.62	4.56	4.53	4.50	4.46	4.43	4.40
6	5.99	5.14	4.76	4.53	4.39	4.28	4.21	4.15	4.10	4.06	4.00	3.94	3.87	3.84	3.81	3.77	3.74	3.70
7	5.59	4.74	4.35	4.12	3.97	3.87	3.79	3.73	3.68	3.64	3.57	3.51	3.44	3.41	3.38	3.34	3.30	3.27
8	5.32	4.46	4.07	3.84	3.69	3.58	3.50	3.44	3.39	3.35	3.28	3.22	3.15	3.12	3.08	3.04	3.01	2.97
9	5.12	4.26	3.86	3.63	3.48	3.37	3.29	3.23	3.18	3.14	3.07	3.01	2.94	2.90	2.86	2.83	2.79	2.75
10	4.96	4.10	3.71	3.48	3.33	3.22	3.14	3.07	3.02	2.98	2.91	2.85	2.77	2.74	2.70	2.66	2.62	2.58
11	4.84	3.98	3.59	3.36	3.20	3.09	3.01	2.95	2.90	2.85	2.79	2.72	2.65	2.61	2.57	2.53	2.49	2.45
12	4.75	3.89	3.49	3.26	3.11	3.00	2.91	2.85	2.80	2.75	2.69	2.62	2.54	2.51	2.47	2.43	2.38	2.34
13	4.67	3.81	3.41	3.18	3.03	2.92	2.83	2.77	2.71	2.67	2.60	2.53	2.46	2.42	2.38	2.34	2.30	2.25
14	4.60	3.74	3.34	3.11	2.96	2.85	2.76	2.70	2.65	2.60	2.53	2.46	2.39	2.35	2.31	2.27	2.22	2.18
15	4.54	3.68	3.29	3.06	2.90	2.79	2.71	2.64	2.59	2.54	2.48	2.40	2.33	2.29	2.25	2.20	2.16	2.11
16	4.49	3.63	3.24	3.01	2.85	2.74	2.66	2.59	2.54	2.49	2.42	2.35	2.28	2.24	2.19	2.15	2.11	2.06
17	4.45	3.59	3.20	2.96	2.81	2.70	2.61	2.55	2.49	2.45	2.38	2.31	2.23	2.19	2.15	2.10	2.06	2.01
18	4.41	3.55	3.16	2.93	2.77	2.66	2.58	2.51	2.46	2.41	2.34	2.27	2.19	2.15	2.11	2.06	2.02	1.97
19	4.38	3.52	3.13	2.90	2.74	2.63	2.54	2.48	2.42	2.38	2.31	2.23	2.16	2.11	2.07	2.03	1.98	1.93
20	4.35	3.49	3.10	2.87	2.71	2.60	2.51	2.45	2.39	2.35	2.28	2.20	2.12	2.08	2.04	1.99	1.95	1.90
21	4.32	3.47	3.07	2.84	2.68	2.57	2.49	2.42	2.37	2.32	2.25	2.18	2.10	2.05	2.01	1.96	1.92	1.87
22	4.30	3.44	3.05	2.82	2.66	2.55	2.46	2.40	2.34	2.30	2.23	2.15	2.07	2.03	1.98	1.94	1.89	1.84
23	4.28	3.42	3.03	2.80	2.64	2.53	2.44	2.37	2.32	2.27	2.20	2.13	2.05	2.01	1.96	1.91	1.86	1.81
24	4.26	3.40	3.01	2.78	2.62	2.51	2.42	2.36	2.30	2.25	2.18	2.11	2.03	1.98	1.94	1.89	1.84	1.79
25	4.24	3.39	2.99	2.76	2.60	2.49	2.40	2.34	2.28	2.24	2.16	2.09	2.01	1.96	1.92	1.87	1.82	1.77
26	4.23	3.37	2.98	2.74	2.59	2.47	2.39	2.32	2.27	2.22	2.15	2.07	1.99	1.95	1.90	1.85	1.80	1.75
27	4.21	3.35	2.96	2.73	2.57	2.46	2.37	2.31	2.25	2.20	2.13	2.06	1.97	1.93	1.88	1.84	1.79	1.73
28	4.20	3.34	2.95	2.71	2.56	2.45	2.36	2.29	2.24	2.19	2.12	2.04	1.96	1.91	1.87	1.82	1.77	1.71
29	4.18	3.33	2.93	2.70	2.55	2.43	2.35	2.28	2.22	2.18	2.10	2.03	1.94	1.90	1.85	1.81	1.75	1.70
30	4.17	3.32	2.92	2.69	2.53	2.42	2.33	2.27	2.21	2.16	2.09	2.01	1.93	1.89	1.84	1.79	1.74	1.68
40	4.08	3.23	2.84	2.61	2.45	2.34	2.25	2.18	2.12	2.08	2.00	1.92	1.84	1.79	1.74	1.69	1.64	1.58
60	4.00	3.15	2.76	2.53	2.37	2.25	2.17	2.10	2.04	1.99	1.92	1.84	1.75	1.70	1.65	1.59	1.53	1.47
120	3.92	3.07	2.68	2.45	2.29	2.18	2.09	2.02	1.96	1.91	1.83	1.75	1.66	1.61	1.55	1.50	1.43	1.35

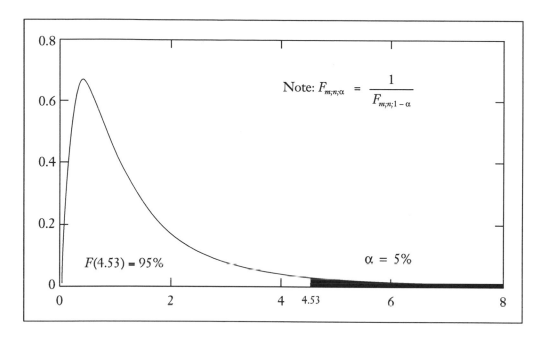

Figure D.1. *F* distribution (*m* = 4, *n* = 6 degrees of freedom).

Table D.3. 97.5th percentile of the *F* distribution.

n	1	2	3	4	5	6	7	8	9	10	12	15	20	24	30	40	60	120
										Numerator (*m*)								
1	647.8	799.5	864.2	899.6	921.8	937.1	948.2	956.6	963.3	968.6	976.7	984.9	993.1	997.3	1001	1006	1010	1014
2	38.51	39.00	39.17	39.25	39.30	39.33	39.36	39.37	39.39	39.40	39.41	39.43	39.45	39.46	39.46	39.47	39.48	39.49
3	17.44	16.04	15.44	15.10	14.88	14.73	14.62	14.54	14.47	14.42	14.34	14.25	14.17	14.12	14.08	14.04	13.99	13.95
4	12.22	10.65	9.98	9.60	9.36	9.20	9.07	8.98	8.90	8.84	8.75	8.66	8.56	8.51	8.46	8.41	8.36	8.31
5	10.01	8.43	7.76	7.39	7.15	6.98	6.85	6.76	6.68	6.62	6.52	6.43	6.33	6.28	6.23	6.18	6.12	6.07
6	8.81	7.26	6.60	6.23	5.99	5.82	5.70	5.60	5.52	5.46	5.37	5.27	5.17	5.12	5.07	5.01	4.96	4.90
7	8.07	6.54	5.89	5.52	5.29	5.12	4.99	4.90	4.82	4.76	4.67	4.57	4.47	4.41	4.36	4.31	4.25	4.20
8	7.57	6.06	5.42	5.05	4.82	4.65	4.53	4.43	4.36	4.30	4.20	4.10	4.00	3.95	3.89	3.84	3.78	3.73
9	7.21	5.71	5.08	4.72	4.48	4.32	4.20	4.10	4.03	3.96	3.87	3.77	3.67	3.61	3.56	3.51	3.45	3.39
10	6.94	5.46	4.83	4.47	4.24	4.07	3.95	3.85	3.78	3.72	3.62	3.52	3.42	3.37	3.31	3.26	3.20	3.14
11	6.72	5.26	4.63	4.28	4.04	3.88	3.76	3.66	3.59	3.53	3.43	3.33	3.23	3.17	3.12	3.06	3.00	2.94
12	6.55	5.10	4.47	4.12	3.89	3.73	3.61	3.51	3.44	3.37	3.28	3.18	3.07	3.02	2.96	2.91	2.85	2.79
13	6.41	4.97	4.35	4.00	3.77	3.60	3.48	3.39	3.31	3.25	3.15	3.05	2.95	2.89	2.84	2.78	2.72	2.66
14	6.30	4.86	4.24	3.89	3.66	3.50	3.38	3.29	3.21	3.15	3.05	2.95	2.84	2.79	2.73	2.67	2.61	2.55
15	6.20	4.77	4.15	3.80	3.58	3.41	3.29	3.20	3.12	3.06	2.96	2.86	2.76	2.70	2.64	2.59	2.52	2.46
16	6.12	4.69	4.08	3.73	3.50	3.34	3.22	3.12	3.05	2.99	2.89	2.79	2.68	2.63	2.57	2.51	2.45	2.38
17	6.04	4.62	4.01	3.66	3.44	3.28	3.16	3.06	2.98	2.92	2.82	2.72	2.62	2.56	2.50	2.44	2.38	2.32
18	5.98	4.56	3.95	3.61	3.38	3.22	3.10	3.01	2.93	2.87	2.77	2.67	2.56	2.50	2.44	2.38	2.32	2.26
19	5.92	4.51	3.90	3.56	3.33	3.17	3.05	2.96	2.88	2.82	2.72	2.62	2.51	2.45	2.39	2.33	2.27	2.20
20	5.87	4.46	3.86	3.51	3.29	3.13	3.01	2.91	2.84	2.77	2.68	2.57	2.46	2.41	2.35	2.29	2.22	2.16
21	5.83	4.42	3.82	3.48	3.25	3.09	2.97	2.87	2.80	2.73	2.64	2.53	2.42	2.37	2.31	2.25	2.18	2.11
22	5.79	4.38	3.78	3.44	3.22	3.05	2.93	2.84	2.76	2.70	2.60	2.50	2.39	2.33	2.27	2.21	2.14	2.08
23	5.75	4.35	3.75	3.41	3.18	3.02	2.90	2.81	2.73	2.67	2.57	2.47	2.36	2.30	2.24	2.18	2.11	2.04
24	5.72	4.32	3.72	3.38	3.15	2.99	2.87	2.78	2.70	2.64	2.54	2.44	2.33	2.27	2.21	2.15	2.08	2.01
25	5.69	4.29	3.69	3.35	3.13	2.97	2.85	2.75	2.68	2.61	2.51	2.41	2.30	2.24	2.18	2.12	2.05	1.98
26	5.66	4.27	3.67	3.33	3.10	2.94	2.82	2.73	2.65	2.59	2.49	2.39	2.28	2.22	2.16	2.09	2.03	1.95
27	5.63	4.24	3.65	3.31	3.08	2.92	2.80	2.71	2.63	2.57	2.47	2.36	2.25	2.19	2.13	2.07	2.00	1.93
28	5.61	4.22	3.63	3.29	3.06	2.90	2.78	2.69	2.61	2.55	2.45	2.34	2.23	2.17	2.11	2.05	1.98	1.91
29	5.59	4.20	3.61	3.27	3.04	2.88	2.76	2.67	2.59	2.53	2.43	2.32	2.21	2.15	2.09	2.03	1.96	1.89
30	5.57	4.18	3.59	3.25	3.03	2.87	2.75	2.65	2.57	2.51	2.41	2.31	2.20	2.14	2.07	2.01	1.94	1.87
40	5.42	4.05	3.46	3.13	2.90	2.74	2.62	2.53	2.45	2.39	2.29	2.18	2.07	2.01	1.94	1.88	1.80	1.72
60	5.29	3.93	3.34	3.01	2.79	2.63	2.51	2.41	2.33	2.27	2.17	2.06	1.94	1.88	1.82	1.74	1.67	1.58
120	5.15	3.80	3.23	2.89	2.67	2.52	2.39	2.30	2.22	2.16	2.05	1.94	1.82	1.76	1.69	1.61	1.53	1.43

Table D.4. 99th percentile of the F distribution.

n	\multicolumn{18}{c}{Numerator (m)}
	1	2	3	4	5	6	7	8	9	10	12	15	20	24	30	40	60	120
1	4052	4999	5404	5624	5764	5859	5928	5981	6022	6056	6107	6157	6209	6234	6260	6286	6313	6340
2	98.50	99.00	99.16	99.25	99.30	99.33	99.36	99.38	99.39	99.40	99.42	99.43	99.45	99.46	99.47	99.48	99.48	99.49
3	34.12	30.82	29.46	28.71	28.24	27.91	27.67	27.49	27.34	27.23	27.05	26.87	26.69	26.60	26.50	26.41	26.32	26.22
4	21.20	18.00	16.69	15.98	15.52	15.21	14.98	14.80	14.66	14.55	14.37	14.20	14.02	13.93	13.84	13.75	13.65	13.56
5	16.26	13.27	12.06	11.39	10.97	10.67	10.46	10.29	10.16	10.05	9.89	9.72	9.55	9.47	9.38	9.29	9.20	9.11
6	13.75	10.92	9.78	9.15	8.75	8.47	8.26	8.10	7.98	7.87	7.72	7.56	7.40	7.31	7.23	7.14	7.06	6.97
7	12.25	9.55	8.45	7.85	7.46	7.19	6.99	6.84	6.72	6.62	6.47	6.31	6.16	6.07	5.99	5.91	5.82	5.74
8	11.26	8.65	7.59	7.01	6.63	6.37	6.18	6.03	5.91	5.81	5.67	5.52	5.36	5.28	5.20	5.12	5.03	4.95
9	10.56	8.02	6.99	6.42	6.06	5.80	5.61	5.47	5.35	5.26	5.11	4.96	4.81	4.73	4.65	4.57	4.48	4.40
10	10.04	7.56	6.55	5.99	5.64	5.39	5.20	5.06	4.94	4.85	4.71	4.56	4.41	4.33	4.25	4.17	4.08	4.00
11	9.65	7.21	6.22	5.67	5.32	5.07	4.89	4.74	4.63	4.54	4.40	4.25	4.10	4.02	3.94	3.86	3.78	3.69
12	9.33	6.93	5.95	5.41	5.06	4.82	4.64	4.50	4.39	4.30	4.16	4.01	3.86	3.78	3.70	3.62	3.54	3.45
13	9.07	6.70	5.74	5.21	4.86	4.62	4.44	4.30	4.19	4.10	3.96	3.82	3.66	3.59	3.51	3.43	3.34	3.25
14	8.86	6.51	5.56	5.04	4.69	4.46	4.28	4.14	4.03	3.94	3.80	3.66	3.51	3.43	3.35	3.27	3.18	3.09
15	8.68	6.36	5.42	4.89	4.56	4.32	4.14	4.00	3.89	3.80	3.67	3.52	3.37	3.29	3.21	3.13	3.05	2.96
16	8.53	6.23	5.29	4.77	4.44	4.20	4.03	3.89	3.78	3.69	3.55	3.41	3.26	3.18	3.10	3.02	2.93	2.84
17	8.40	6.11	5.19	4.67	4.34	4.10	3.93	3.79	3.68	3.59	3.46	3.31	3.16	3.08	3.00	2.92	2.83	2.75
18	8.29	6.01	5.09	4.58	4.25	4.01	3.84	3.71	3.60	3.51	3.37	3.23	3.08	3.00	2.92	2.84	2.75	2.66
19	8.18	5.93	5.01	4.50	4.17	3.94	3.77	3.63	3.52	3.43	3.30	3.15	3.00	2.92	2.84	2.76	2.67	2.58
20	8.10	5.85	4.94	4.43	4.10	3.87	3.70	3.56	3.46	3.37	3.23	3.09	2.94	2.86	2.78	2.69	2.61	2.52
21	8.02	5.78	4.87	4.37	4.04	3.81	3.64	3.51	3.40	3.31	3.17	3.03	2.88	2.80	2.72	2.64	2.55	2.46
22	7.95	5.72	4.82	4.31	3.99	3.76	3.59	3.45	3.35	3.26	3.12	2.98	2.83	2.75	2.67	2.58	2.50	2.40
23	7.88	5.66	4.76	4.26	3.94	3.71	3.54	3.41	3.30	3.21	3.07	2.93	2.78	2.70	2.62	2.54	2.45	2.35
24	7.82	5.61	4.72	4.22	3.90	3.67	3.50	3.36	3.26	3.17	3.03	2.89	2.74	2.66	2.58	2.49	2.40	2.31
25	7.77	5.57	4.68	4.18	3.85	3.63	3.46	3.32	3.22	3.13	2.99	2.85	2.70	2.62	2.54	2.45	2.36	2.27
26	7.72	5.53	4.64	4.14	3.82	3.59	3.42	3.29	3.18	3.09	2.96	2.81	2.66	2.58	2.50	2.42	2.33	2.23
27	7.68	5.49	4.60	4.11	3.78	3.56	3.39	3.26	3.15	3.06	2.93	2.78	2.63	2.55	2.47	2.38	2.29	2.20
28	7.64	5.45	4.57	4.07	3.75	3.53	3.36	3.23	3.12	3.03	2.90	2.75	2.60	2.52	2.44	2.35	2.26	2.17
29	7.60	5.42	4.54	4.04	3.73	3.50	3.33	3.20	3.09	3.00	2.87	2.73	2.57	2.49	2.41	2.33	2.23	2.14
30	7.56	5.39	4.51	4.02	3.70	3.47	3.30	3.17	3.07	2.98	2.84	2.70	2.55	2.47	2.39	2.30	2.21	2.11
40	7.31	5.18	4.31	3.83	3.51	3.29	3.12	2.99	2.89	2.80	2.66	2.52	2.37	2.29	2.20	2.11	2.02	1.92
60	7.08	4.98	4.13	3.65	3.34	3.12	2.95	2.82	2.72	2.63	2.50	2.35	2.20	2.12	2.03	1.94	1.84	1.73
120	6.85	4.79	3.95	3.48	3.17	2.96	2.79	2.66	2.56	2.47	2.34	2.19	2.03	1.95	1.86	1.76	1.66	1.53

Table D.5. 99.5th percentile of the *F* distribution.

n	Numerator (*m*)																	
	1	2	3	4	5	6	7	8	9	10	12	15	20	24	30	40	60	120
1	16212	19997	21614	22501	23056	23440	23715	23924	24091	24222	24427	24632	24837	24937	25041	25146	25254	25358
2	198.5	199.0	199.2	199.2	199.3	199.3	199.4	199.4	199.4	199.4	199.4	199.4	199.4	199.4	199.5	199.5	199.5	199.5
3	55.55	49.80	47.47	46.20	45.39	44.84	44.43	44.13	43.88	43.68	43.39	43.08	42.78	42.62	42.47	42.31	42.15	41.99
4	31.33	26.28	24.26	23.15	22.46	21.98	21.62	21.35	21.14	20.97	20.70	20.44	20.17	20.03	19.89	19.75	19.61	19.47
5	22.78	18.31	16.53	15.56	14.94	14.51	14.20	13.96	13.77	13.62	13.38	13.15	12.90	12.78	12.66	12.53	12.40	12.27
6	18.63	14.54	12.92	12.03	11.46	11.07	10.79	10.57	10.39	10.25	10.03	9.81	9.59	9.47	9.36	9.24	9.12	9.00
7	16.24	12.40	10.88	10.05	9.52	9.16	8.89	8.68	8.51	8.38	8.18	7.97	7.75	7.64	7.53	7.42	7.31	7.19
8	14.69	11.04	9.60	8.81	8.30	7.95	7.69	7.50	7.34	7.21	7.01	6.81	6.61	6.50	6.40	6.29	6.18	6.06
9	13.61	10.11	8.72	7.96	7.47	7.13	6.88	6.69	6.54	6.42	6.23	6.03	5.83	5.73	5.62	5.52	5.41	5.30
10	12.83	9.43	8.08	7.34	6.87	6.54	6.30	6.12	5.97	5.85	5.66	5.47	5.27	5.17	5.07	4.97	4.86	4.75
11	12.23	8.91	7.60	6.88	6.42	6.10	5.86	5.68	5.54	5.42	5.24	5.05	4.86	4.76	4.65	4.55	4.45	4.34
12	11.75	8.51	7.23	6.52	6.07	5.76	5.52	5.35	5.20	5.09	4.91	4.72	4.53	4.43	4.33	4.23	4.12	4.01
13	11.37	8.19	6.93	6.23	5.79	5.48	5.25	5.08	4.94	4.82	4.64	4.46	4.27	4.17	4.07	3.97	3.87	3.76
14	11.06	7.92	6.68	6.00	5.56	5.26	5.03	4.86	4.72	4.60	4.43	4.25	4.06	3.96	3.86	3.76	3.66	3.55
15	10.80	7.70	6.48	5.80	5.37	5.07	4.85	4.67	4.54	4.42	4.25	4.07	3.88	3.79	3.69	3.59	3.48	3.37
16	10.58	7.51	6.30	5.64	5.21	4.91	4.69	4.52	4.38	4.27	4.10	3.92	3.73	3.64	3.54	3.44	3.33	3.22
17	10.38	7.35	6.16	5.50	5.07	4.78	4.56	4.39	4.25	4.14	3.97	3.79	3.61	3.51	3.41	3.31	3.21	3.10
18	10.22	7.21	6.03	5.37	4.96	4.66	4.44	4.28	4.14	4.03	3.86	3.68	3.50	3.40	3.30	3.20	3.10	2.99
19	10.07	7.09	5.92	5.27	4.85	4.56	4.34	4.18	4.04	3.93	3.76	3.59	3.40	3.31	3.21	3.11	3.00	2.89
20	9.94	6.99	5.82	5.17	4.76	4.47	4.26	4.09	3.96	3.85	3.68	3.50	3.32	3.22	3.12	3.02	2.92	2.81
21	9.83	6.89	5.73	5.09	4.68	4.39	4.18	4.01	3.88	3.77	3.60	3.43	3.24	3.15	3.05	2.95	2.84	2.73
22	9.73	6.81	5.65	5.02	4.61	4.32	4.11	3.94	3.81	3.70	3.54	3.36	3.18	3.08	2.98	2.88	2.77	2.66
23	9.63	6.73	5.58	4.95	4.54	4.26	4.05	3.88	3.75	3.64	3.47	3.30	3.12	3.02	2.92	2.82	2.71	2.60
24	9.55	6.66	5.52	4.89	4.49	4.20	3.99	3.83	3.69	3.59	3.42	3.25	3.06	2.97	2.87	2.77	2.66	2.55
25	9.48	6.60	5.46	4.84	4.43	4.15	3.94	3.78	3.64	3.54	3.37	3.20	3.01	2.92	2.82	2.72	2.61	2.50
26	9.41	6.54	5.41	4.79	4.38	4.10	3.89	3.73	3.60	3.49	3.33	3.15	2.97	2.87	2.77	2.67	2.56	2.45
27	9.34	6.49	5.36	4.74	4.34	4.06	3.85	3.69	3.56	3.45	3.28	3.11	2.93	2.83	2.73	2.63	2.52	2.41
28	9.28	6.44	5.32	4.70	4.30	4.02	3.81	3.65	3.52	3.41	3.25	3.07	2.89	2.79	2.69	2.59	2.48	2.37
29	9.23	6.40	5.28	4.66	4.26	3.98	3.77	3.61	3.48	3.38	3.21	3.04	2.86	2.76	2.66	2.56	2.45	2.33
30	9.18	6.35	5.24	4.62	4.23	3.95	3.74	3.58	3.45	3.34	3.18	3.01	2.82	2.73	2.63	2.52	2.42	2.30
40	8.83	6.07	4.98	4.37	3.99	3.71	3.51	3.35	3.22	3.12	2.95	2.78	2.60	2.50	2.40	2.30	2.18	2.06
60	8.49	5.79	4.73	4.14	3.76	3.49	3.29	3.13	3.01	2.90	2.74	2.57	2.39	2.29	2.19	2.08	1.96	1.83
120	8.18	5.54	4.50	3.92	3.55	3.28	3.09	2.93	2.81	2.71	2.54	2.37	2.19	2.09	1.98	1.87	1.75	1.61

Control Chart Formulas and Control Chart Factors

Control limits (LCL, UCL) for samples of n measurements

	Estimate for process variation (Given)		
	Actual standard deviation, σ	Average standard deviation, $\bar{s}$, for sample of n	Average range, $\bar{R}$, for sample of n
Chart for process mean (centerline is $\bar{\bar{x}}$ or μ).	$\mu \pm 3\dfrac{\sigma}{\sqrt{n}}$	$\bar{\bar{x}} \pm A_3\bar{s}$	$\bar{\bar{x}} \pm A_2\bar{R}$
Chart for process variation	s chart: $[B_5\sigma, B_6\sigma]$ Centerline, $c_4\sigma$ R chart: $[D_1\sigma, D_2\sigma]$ Centerline, $d_2\sigma$	s chart: $[B_3\bar{s}, B_4\bar{s}]$ Centerline, $\bar{s}$	R chart: $[D_3\bar{R}, D_4\bar{R}]$ Centerline, $\bar{R}$
Estimate for the process standard deviation σ	Given	For m samples, where the ith sample size is n_i, and $c_4(n)$ is c_4 for a sample of n, $\hat{\sigma} = \dfrac{1}{m}\,\Sigma_{i=1}^{m}\dfrac{s_i}{c_4(n_i)}$	For m samples, where the ith sample size is n_i, and $d_2(n)$ is d_2 for a sample of n, $\hat{\sigma} = \dfrac{1}{m}\,\Sigma_{i=1}^{m}\dfrac{R_i}{d_2(n_i)}$

Control Chart Factors

	Factors for $\bar{x}$ chart			Factors for central line (s charts)		Factors for s chart				Factors for central line (range charts)			Factors for R chart			
						Given $\bar{s}$		Given σ					Given σ		Given $\bar{R}$	
n	A	A_2	A_3	c_4	$1/c_4$	B_3	B_4	B_5	B_6	d_2	d_3	$1/d_2$	D_1	D_2	D_3	D_4
2	2.121	1.880	2.659	0.7979	1.2533	0.000	3.267	0.000	2.606	1.128	0.853	0.8862	0.000	3.686	0.000	3.267
3	1.732	1.023	1.954	0.8862	1.1284	0.000	2.568	0.000	2.276	1.693	0.888	0.5908	0.000	4.358	0.000	2.575
4	1.500	0.729	1.628	0.9213	1.0854	0.000	2.266	0.000	2.088	2.059	0.880	0.4857	0.000	4.698	0.000	2.282
5	1.342	0.577	1.427	0.9400	1.0638	0.000	2.089	0.000	1.964	2.326	0.864	0.4299	0.000	4.918	0.000	2.114
6	1.225	0.483	1.287	0.9515	1.0509	0.030	1.970	0.029	1.874	2.534	0.848	0.3946	0.000	5.079	0.000	2.004
7	1.134	0.419	1.182	0.9594	1.0424	0.118	1.882	0.113	1.806	2.704	0.833	0.3698	0.205	5.204	0.076	1.924
8	1.061	0.373	1.099	0.9650	1.0362	0.185	1.815	0.179	1.751	2.847	0.820	0.3512	0.388	5.307	0.136	1.864
9	1.000	0.337	1.032	0.9693	1.0317	0.239	1.761	0.232	1.707	2.970	0.808	0.3367	0.547	5.394	0.184	1.816
10	0.949	0.308	0.975	0.9727	1.0281	0.284	1.716	0.276	1.669	3.078	0.797	0.3249	0.686	5.469	0.223	1.777
11	0.905	0.285	0.927	0.9754	1.0253	0.321	1.679	0.313	1.637	3.173	0.787	0.3152	0.811	5.535	0.256	1.744
12	0.866	0.266	0.886	0.9776	1.0230	0.354	1.646	0.346	1.610	3.258	0.778	0.3069	0.923	5.594	0.283	1.717
13	0.832	0.249	0.850	0.9794	1.0210	0.382	1.618	0.374	1.585	3.336	0.770	0.2998	1.025	5.647	0.307	1.693
14	0.802	0.235	0.817	0.9810	1.0194	0.406	1.594	0.398	1.563	3.407	0.763	0.2935	1.118	5.696	0.328	1.672
15	0.775	0.223	0.789	0.9823	1.0180	0.428	1.572	0.421	1.544	3.472	0.756	0.2880	1.203	5.740	0.347	1.653
16	0.750	0.212	0.763	0.9835	1.0168	0.448	1.552	0.440	1.527	3.532	0.750	0.2831	1.282	5.782	0.363	1.637
17	0.728	0.203	0.739	0.9845	1.0157	0.466	1.534	0.459	1.510	3.588	0.744	0.2787	1.356	5.820	0.378	1.622
18	0.707	0.194	0.718	0.9854	1.0148	0.482	1.518	0.475	1.496	3.640	0.739	0.2747	1.424	5.856	0.391	1.609
19	0.688	0.187	0.698	0.9862	1.0140	0.497	1.503	0.490	1.483	3.689	0.733	0.2711	1.489	5.889	0.404	1.596
20	0.671	0.180	0.680	0.9869	1.0132	0.510	1.490	0.503	1.470	3.735	0.729	0.2677	1.549	5.921	0.415	1.585
21	0.655	0.173	0.663	0.9876	1.0126	0.523	1.477	0.516	1.459	3.778	0.724	0.2647	1.606	5.951	0.425	1.575
22	0.640	0.167	0.647	0.9882	1.0120	0.534	1.466	0.528	1.448	3.819	0.720	0.2618	1.660	5.979	0.435	1.565
23	0.626	0.162	0.633	0.9887	1.0114	0.545	1.455	0.539	1.438	3.858	0.716	0.2592	1.711	6.006	0.443	1.557
24	0.612	0.157	0.619	0.9892	1.0109	0.555	1.445	0.549	1.429	3.895	0.712	0.2567	1.759	6.032	0.452	1.548
25	0.600	0.153	0.606	0.9896	1.0105	0.565	1.435	0.559	1.420	3.931	0.708	0.2544	1.805	6.056	0.459	1.541

Formulas for Control Chart Factors*

These are needed only if it is necessary to program a computer to calculate the factors.

$$c_4 = \sqrt{\frac{2}{n-1}} \frac{\left(\frac{n-2}{2}\right)!}{\left(\frac{n-3}{2}\right)!} = \sqrt{\frac{2}{n-1}} \frac{\Gamma\left(\frac{n}{2}\right)}{\Gamma\left(\frac{n-1}{2}\right)}$$	$$B_5 = c_4 - 3\sqrt{1 - c_4^2}$$ $$B_6 = c_4 + 3\sqrt{1 - c_4^2}$$
$$A = \frac{3}{\sqrt{n}}$$	$$B_3 = \frac{B_5}{c_4} \quad B_4 = \frac{B_6}{c_4}$$
$$A_2 = \frac{3}{d_2\sqrt{n}}$$	$$D_1 = d_2 - 3d_3$$ $$D_1 = d_2 + 3d_3$$ See below for d_2, d_3
$$A_3 = \frac{3}{c_4\sqrt{n}}$$	$$D_3 = \frac{D_1}{d_2} \quad D_4 = \frac{D_2}{d_2}$$

The formulas for d_2 and d_3 are computation intensive, but a 486 or Pentium processor should be adequate to reproduce their results.

$$d_2 = \int_{-\infty}^{\infty}\left[1 - (1 - \Phi(x))^n - \Phi(x)^n\right]dx \quad \text{where } \Phi(x) \text{ is the cumulative normal function.}$$

$$d_3 = \left[2\int_{-\infty}^{\infty}\int_{-\infty}^{x_1}\left[1 - \Phi(x_1)^n - (1 - \Phi(x_n))^n + (\Phi(x_1) - \Phi(x_n))^n\right]dx_n dx_1 - d_2^2\right]^{\frac{1}{2}}$$

$\infty = 5$ was adequate for MathCAD to reproduce tabulated results (ASTM 1990, Table 49)

---

*See ASTM 1990, 91–95.